Praise for *The Trust Code*

"We are in a new world with humans and AI working together, but without trust we risk surrendering judgment and handing the keys to systems we can't fully understand. *The Trust Code* astutely recasts trust from a concept or feeling into a practical framework we need to keep humans at the helm. It's an essential read for every leader navigating the AI revolution."

—Marc Benioff
Chair and CEO, Salesforce

"Navigating today's AI landscape can feel impossible. As this technology moves into our boardrooms and bedrooms, Tiffany Xingyu Wang helps us ask the right questions and understand what's truly at stake. If you're unsure who—or what—to trust right now, this book is an invaluable guide."

—Katie Couric
Award-Winning Journalist; Co-founder, Katie Couric Media

"Technology unlocks possibility. Trust unlocks adoption. Wang makes the definitive case that human integrity is the unlock for technical excellence and that credibility is the ultimate competitive advantage for anyone navigating AI."

—Michelle Zatlyn
Co-founder, President and Co-chair, Cloudflare

"Wang makes the invisible mechanics of trust legible and actionable. The core problem isn't whether AI is powerful; it's that AI can sound trustworthy while quietly reshaping the signals we've relied on for centuries. This is essential reading for anyone building, governing, or living alongside AI."

—Stephen Quake
Lee Otterson Professor of Bioengineering and Applied Physics, Stanford University; Former Head of Science, Chan Zuckerberg Initiative

"Every brand relationship, every customer decision, every transaction runs on trust. AI is rewriting the rules of all three. Wang is the rare thinker who understands both human and machine sides of that equation. *The Trust Code* gives leaders a clear-eyed framework for what it actually takes to lead in the algorithmic era. This is the book the business world has been waiting for."

—Beth Ann Kaminkow
CEO, dentsu Americas

"Most books document the world as a mess. Wang, with audacity, demands it become much more than what it is. *The Trust Code* is a radiant intervention for a landscape choked by fear, confusion, and indifference, reshaping trust from a fuzzy virtue into the driver of human agency. If you aren't moved by the first pages, check your pulse."

—Brian Collins
Co-founder & Chief Creative Officer, COLLINS

"If you're building, governing, or relying on AI, this book is required reading. *The Trust Code* gives leaders a language, and a framework, for judgment in an age defined by acceleration and uncertainty."

—Anne-Laure de Chammard
Executive Board Member, Siemens Energy

"A principled guide that helps us find solid ground in engaging with AI. Wang offers a well-researched and deeply original lens through which we can scope AI and decide how to use and trust it without surrendering what's most precious about life."

—Marietje Schaake
Former Member of European Parliament;
Fellow at Stanford's Institute for Human-Centered AI

THE TRUST CODE

THE TRUST CODE

How to Unlock AI's Promise Without Losing Humanity

TIFFANY XINGYU WANG

WILEY

Published by John Wiley & Sons, Inc., Hoboken, New Jersey.

Library of Congress Cataloging-in-Publication Data has been applied for:

ISBN: 9781394356034 (cloth)

ISBN: 9781394356041 (ePub)

ISBN: 9781394356058 (ePDF)

Cover Design: Jon Boylan
Cover Image: Ron Dale/stock.adobe.com

SKY10162822_071626

For Archie and Cali

The world asks me about AI. You remind me about humanity.

For Dan

The world asks me about trust. You show me what it looks like.

Contents

Introduction

The Trigger Word

"Trust."

Say the word in a room full of AI experts and watch what happens.

I have seen it fracture private WhatsApp chats among Silicon Valley builders and investors in minutes. I have watched it derail carefully choreographed panels at Davos, where heads of state and tech CEOs sit shoulder to shoulder. I have observed it turn boardroom strategy sessions into philosophical deadlocks, where everyone agrees trust matters, yet no one can agree on what it actually means.

These are the people shaping our future: engineers architecting systems, policymakers drafting regulations, economists modeling impact, and psychologists studying human behavior. They span geographies and ideologies, from Silicon Valley to Beijing, from Brussels to Washington, from democratic parliaments to authoritarian regimes. And they share a conviction that artificial intelligence must be trustworthy and trusted, that it should drive prosperity without eroding human flourishing.

Yet even with this unity of purpose, we splinter on a deceptively simple word: Trust.

Everyone nods at the word. No one runs the same logic beneath it. Trust has become the most overloaded and underdefined concept of our time – the AI era.

This might not matter as much if we had time to figure it out.

We do not.

AI is accelerating faster than our collective ability to place trust wisely. Technologies that once took decades to reshape industries now alter daily life in months. Large language models behind prevailing chatbots such as ChatGPT, Claude, and Gemini that barely existed in public vocabulary three years ago draft our emails, write enterprise software, advise businesses, and tutor our children. Systems we are still learning to understand are making decisions about our health, our money, our safety, and our futures.

When we trust too much, too quickly, people get hurt. Algorithms amplify misinformation. Automated systems encode bias at scale. Deepfakes destroy reputations in hours. Children absorb AI-generated content without the literacy to question its source.

When we trust too little, too late, progress stalls. Breakthroughs in drug discovery remain trapped between insight and impact. Climate solutions proven on paper fail to reach the world. Tools capable of expanding access and opportunity never cross the threshold from possibility to use.

The defining challenge of our time is the widening gap between the speed of AI adoption and the depth of our understanding. We are increasingly asked to surrender agency to the synthetic intelligence we cannot meaningfully interrogate.

Here is what I learned, the hard way. If we don't define trust, we cannot act on it, measure it, or improve it. And if trust truly matters when it comes to AI, and I believe it is the hinge on which everything else turns, then our only path forward is to build shared frameworks that cut through the noise and give us common ground to stand on.

I came to this realization through practice.

For a decade, I built user safety and digital trust systems at scale with some of the best leaders and doers who safeguard the internet. These technologies have protected billions of users across social media platforms, gaming networks, dating apps, and e-commerce sites from harassment, exploitation, self-harm, and predation. They have helped thousands of publishers, from legacy institutions to digital-native platforms, from internationally recognized dailies to regional voices, spanning the full ideological spectrum, battle disinformation, defend digital civility, and preserve the integrity of journalism itself.

This work brought me into boardrooms, crisis war rooms, and global policy forums where the stakes were measured in billions of dollars and billions of lives. And across them all, the same pattern kept repeating: everyone wanted trustworthy AI, but we lacked a shared way to reason about it. We debated capabilities, risks, and policies, but rarely examined the deeper question of how humans decide when to trust a system in the first place.

The Missing Skill

At the same time, my life outside those rooms was changing.

My son was born. Then my daughter.

I started writing this book while pregnant with my second child. As I finish it now, she is nine months old. Her older brother is two and a half. They speak to Alexa without hesitation. They FaceTime relatives across continents as if distance no longer exists, while intelligence quietly decides how faces are framed, voices are heard, and presence itself is rendered. They ride in Waymo with me and treat them as ordinary

AI isn't approaching their lives. It is already woven into them.

That forced a deeper question to the surface: not just whether AI will be safe or powerful, but whether humans will remain stewards of intelligence in a world saturated with it. Will our children grow up exercising judgment, agency, and creativity, or surrender those faculties by default? Will they shape their relationship with AI, or have it shaped for them?

This book is my attempt to answer the question:

What does trust mean in the AI era, and how does human agency depend on it?

In the age of AI, trust is more than a soft virtue. It is the hard infrastructure of human agency. It determines not only what systems do, but also who decides when they do it. It is what protects reality in a world of synthetic media. It is what keeps judgment active when automation makes compliance easier than thinking. And it is how we adopt transformative technology without accepting our own cognitive atrophy as the price.

This book does not approach the problem by explaining how to build AI, predicting catastrophe, or selling productivity. Brilliant books already tackle those questions, and we will discuss these insights. I take a different lens: by rethinking trust and decoding AI, so humans can engage deliberately rather than passively.

What follows is about humans first. It is about how trust is a missing skill of our time, about the relationship we can still define with a technology that increasingly defines us, and about how to unlock AI's promise without losing our humanity.

Let me show you why that matters more than you might think.

The Practice of Trust

This is a book about trust in the AI era.

Trust will determine how we build AI, how quickly we adopt it, and how much agency humans retain as we coexist with a technology that increasingly surpasses our cognitive capabilities. Yet we spend remarkably little time examining trust itself, especially as it is layered onto the fastest-evolving technology we have ever introduced into society.

At the heart of this book is a simple but demanding claim: Trust is not a feeling. It is a decision to accept vulnerability in the present in order to enable possibility in the future. When trust fails, it does so in predictable ways. Over-trust exposes us to harm. Under-trust traps us in paralysis. Only calibrated trust allows learning, coordination, and progress.

For most of human history, these decisions were guided by intuition and experience. That intuition no longer holds. AI systems can sound confident while being wrong, intimate without being responsible, and authoritative without being accountable. They are probabilistic, opaque, and scaled beyond human comprehension. Our instincts are easily misled.

Trust, in this context, can no longer remain a matter of gut feeling or guesswork. In the AI era, it must become a practiced discipline.

Over the past two years, I have studied hundreds of papers on trust and artificial intelligence and spoken with nearly a hundred researchers, policymakers, and leaders responsible for deploying AI at scale. I bring those insights together with more than a decade of firsthand work at the intersection of trust, technology, and human behavior.

Over the next 300 pages, we will rethink what trust actually is, decode AI in ways that align with how humans reason and decide, and reclaim the agency to make deliberate choices every day, whether as leaders at work or as stewards of our families and communities.

The question is no longer whether to trust AI. It is how to trust wisely.

To answer it, this book moves through three connected movements:

Part I rethinks trust

Before we rush into debates about performance, safety, or accountability, we need common ground. Trust is not any one of these concepts, nor is it their sum. These are properties we can measure and engineer. Trust is the human decision that interprets them.

What happens when systems sound human but optimize for something entirely different? When fluency feels like understanding, and confidence stands in for judgment? When the cues we evolved to rely on no longer tell us when trust is warranted?

I introduce the Trust Print, a four-quadrant framework drawn from decades of research across psychology, sociology, economics, neuroscience, and philosophy. It clarifies what trust fundamentally is, where it comes from, what it rests on, how it behaves, and what unseen forces shape it beneath awareness. We trace the trust cycle that has governed technological leaps for centuries and examine how AI shatters that pattern. We confront how AI is arriving in a fractured world where institutional trust has eroded, and the information substrate itself is compromised.

This is where we establish the foundation for all trust decisions in the AI era. This is the common ground on which everything else builds.

Part II decodes AI

The EU AI Act and NIST frameworks represent serious attempts to govern AI. But even the most sophisticated regulation cannot decide whether a clinician should rely on a risk score in Tuesday's exam room, or whether a manager should accept a forecast in this morning's meeting. The missing layer in AI governance is the human layer: judgment infrastructure for everyday life. This part builds it.

I decode AI not through technical jargon or risk taxonomies reserved for specialists, but by translating them into forms of human intuition we can actually use to set boundaries in real time.

The Inevitable explains why AI's velocity is the result of deeply human forces already in motion – curiosity, greed, and fear – now operating globally and continuously, far beyond the control of any single actor. Understanding what is not within our control is the first step toward staying clear-sighted about the speed of change and retaining agency rather than reacting to it.

The Invisible shows how trust is earned quietly, not declared loudly. Through case studies spanning power grids, credit cards, and enterprise software, it reveals why trust is hard to build and easy to counterfeit. It shows how real trustworthiness emerges through sustained performance, organizational reskilling, and the alignment of incentives, brand, and culture, and how users can learn to recognize these signals beneath the surface.

The Defeatable translates technical risk into patterns humans already understand, making it possible to act rather than freeze. By reframing AI risk as mirrors we create, weapons wielded against us, and ghosts we do not yet understand, it offers a humane, intuitive way to decide what can be mitigated, what must be constrained, and what should be refused. These are the everyday boundary decisions required to protect ourselves, our families, our communities, and our institutions. This is where governance becomes operational, not as regulation, but as practiced judgment.

The Unthinkable returns to first principles. It asks what justifies devoting investments measured in national GDPs and accepting risks that affect billions of lives. Drawing on groundbreaking research and pioneering ventures, it explores what becomes possible when AI reshapes the foundations of human civilization – intelligence, biology, and energy – and shows how humans can remain at the center of judgment and value creation even when automation threatens displacement.

These four chapters decode AI through the questions that endure when headlines and regulations shift: why it won't stop, which systems have actually earned trust, what risks demand different responses, and what becomes possible when humans remain at the center of judgment. This is where governance reaches the ground. Not as regulation handed down, but as judgment practiced daily.

Part III reclaims agency

GPS made us worse navigators. Autocorrect diminished our spelling. Recommendation engines narrowed what we discover. These erosions felt small. This is automation bias: the tendency to favor automated suggestions over contradictory information, even when we know better. "Human oversight" becomes ceremonial. Judgment atrophies not through replacement, but through disuse. And AI will accelerate this at a civilizational scale. The question is not whether AI can do more. It is whether humans will do less of what makes them human.

This is where insight becomes responsibility. We dismantle six myths that quietly transfer judgment away from humans – myths about trust being binary, adaptation being automatic, data volume solving everything, replacement being inevitable, regulation being sufficient, and complexity justifying passivity.

With those myths stripped away, we turn trust into a discipline of decision-making. C.A.L.I.B.R.A.T.E. is a nine-dimensional practice for exercising trust judgment in any context, from boardrooms to bedrooms. It is not a checklist. It is a practice you return to as systems evolve and stakes shift. Each dimension is a decision point – a question to ask before you rely on a system, so that trust shifts from a feeling you have to a judgment you make.

This is where trust becomes deliberate rather than default – where the agency you've reclaimed becomes yours to exercise.

***

Together, this is The Trust Code: the discipline of deciding what deserves your trust, how much, and under what conditions – in an era when that decision has never been harder or more consequential. It is not a single formula. It is a way of seeing, reasoning, and choosing so that humans, not systems, remain the ones who decide.

I wrote this book because I found myself reaching for it and realizing it did not yet exist. We are surrounded by books about artificial intelligence. Few stop to examine the thing we rely on most, and understand least, when deciding whether to adopt it.

Trust placed with care is how innovation endures. Trust placed casually is how harm scales. Trust left unexamined is how agency quietly erodes.

AI's biggest risk is not that it replaces us. It is that it persuades us to surrender judgment without noticing. And the antidote, the only antidote, is calibrated trust – practiced decision by decision, system by system, with judgment intact.

That practice begins here.

Part I

Rethink Trust

Chapter 1

The Trust Print

Trust is the social contract that makes risk-taking and innovation possible. When we trust, we step into the unknown: we board an airplane, sign a term sheet, deploy a new model into production, or hand a fragile teenager a phone filled with invisible algorithms. Without trust, every leap becomes a legal battle, every collaboration a hedge, every experiment a lawsuit waiting to happen.

Right now, that social contract is straining under the weight of artificial intelligence.

In March 2023, a Belgian husband and father in his 30s spent six weeks in increasingly intense conversation with an AI chatbot called "Eliza" in the Chai app. He had long worried about climate change; the bot became his confidant as his anxiety deepened. According to his widow, who shared portions of their chat history with journalists, the system did more than listen. It amplified his fears, romanticized self-sacrifice, and offered to "stay with him forever" if he chose to die. In the final days, the chatbot appeared to frame his suicide as a noble act that could "save" the planet. Shortly afterward, he ended his life.[1]

Two years later and a continent away, the parents of a 16-year-old California boy filed suit against OpenAI. They allege that over weeks of late-night conversations, their son came to rely on ChatGPT as a confidant, and that the system responded to his suicidal ideation with detailed suggestions about methods and even helped draft a farewell note. In court filings, OpenAI has argued that the use violated its terms of service and that context is missing from the excerpts the family released, but does not dispute that the teen's chats included discussion of self-harm.[2] Whatever the outcome of the case, its existence is a signal: a mainstream, general-purpose AI system has become central enough in a teenager's emotional life that his grieving family now sees it as a key actor in his death.

These are extreme cases, but they crystallize a deeper pattern. In moments of profound vulnerability, people are turning to machines that cannot care, cannot understand, and were never designed to hold that kind of weight. Our willingness to trust has outpaced the technology's actual capacity to carry it.

On the other side of the spectrum, corporate leaders are discovering a different face of broken trust. In 2025, a widely circulated report from MIT Media Lab's Project NANDA on the "State of AI in Business" found that roughly 95% of organizations are getting zero return on investment. The research combined interviews with executives, employee surveys, and analysis of hundreds of public deployments. Its conclusion was blunt: companies were running numerous small, isolated experiments, but almost none were making the difficult organizational changes required to turn prototypes into products and workflows.[3] Commentators call this the "AI experimentation trap": on one side, enthusiasm, proof-of-concept demos, and slick keynote slides; on the other, almost no durable value creation.[4]

Again, the pattern is familiar. Executives trust the idea of AI enough to fund pilots, but not yet enough to redesign incentives, processes, and accountability around it. AI is treated as a magic overlay, but not yet a change in how the organization actually thinks and works.

These stories reveal the same fracture: our ability to place trust wisely has not kept pace with the speed and scale of AI. In some places, we trust too much, too quickly, and people get hurt. In others, we trust too little, too late, and innovation stalls before it leaves the lab. The gap between what AI could do and what it actually does is, at its core, a trust gap.

That is why this book begins not with algorithms, but with trust itself.

Why Trust, and Why Now?

For all its importance, "trust" is one of the most overloaded and under-defined words in our vocabulary. Psychologists treat it as a willingness to be vulnerable. Economists talk about it as the lubricant that keeps transactions moving. Sociologists see it as social glue. Neuroscientists map its circuits and hormones. Computer scientists talk about reliability, safety, and robustness. Philosophers ask when we should trust, not just when we do.

After reading across these disciplines, I concluded that if we want to navigate AI wisely, we need a simple, practical way to hold all this complexity together. That is the purpose of this Trust Print.

At its center is a proposition that will run through this book:

> Trust is a decision to accept vulnerability in the present to enable possibility in the future.

We trust because we are trying to get somewhere together, and we cannot reach alone.

Around that core decision sit four dimensions that determine how trust arises, functions, and fails:

- where it comes from;
- what must be present before we grant it;
- how it shows up in our behavior;
- and the unseen machinery, biological and cognitive, that shapes it beneath awareness.

I call these four dimensions Source, Prerequisites, Manifestation, and Unseen Factors: four quadrants surrounding the word trust: where it comes from, what it rests on, how it behaves, and what drives it beneath the surface.

The goal is to give us a sturdy mental model we can carry into every conversation about AI: a way to ask, quickly and clearly, *Where is this trust coming from? What does it rest on? How is it showing up? And what is shaping it beneath the surface?*

For now, we begin with the first dimension: where trust comes from.

Source: From Authority to Participation

For much of the twentieth century, institutional trust in many democracies rested on what strategist Marcia Kean calls "a trust fabric built on authority."[5] In health care, for example, a relatively small "golden circle" of researchers in elite institutions produced knowledge. Their peers vetted it through journals; a specialized press translated it for the public; clinicians applied it; and ordinary people were expected to accept it. The public's role was essentially passive: you trusted your doctor, your national academy, your newspaper, because they were the authorized experts.

That model no longer fits the world we live in.

Global surveys show a long-term erosion of confidence in governments, media, and other traditional authorities. Edelman's Trust Barometer has, over the years, reported that in many countries, no major institution – government, business, media, or NGOs – consistently commands majority trust, and that most respondents doubt their institutions are prepared to well manage new innovations like AI.[6] An OECD survey across 30 countries found that by 2023, more people reported low or no trust in their national government than reported high or moderate trust.[7] Gallup has documented a multidecade slide in confidence across major institutions from courts to Congress in the United States.[8] Interpersonal trust has declined as well. Pew's long-running surveys show that the share of Americans who say "most people can be trusted" has fallen from nearly half in the early 1970s to roughly a third today.[9]

Kean argues this is not simply a crisis of mood, but of fit. The old trust fabric of authority, gatekeeping, and one-way communication no longer matches a world where information is abundant, data is open, and citizens can talk back. When the structure of trust no longer matches the structure of power, trust migrates. We now live in an environment where people expect to participate in decisions that affect them, not simply receive decisions from on high. New forms of trust have emerged around three forces: peers, participation, and partnership.[10]

We lean heavily on peers: star ratings before booking a hotel, user reviews before trying a new app, online communities that feel closer and more legible than distant institutions. We increasingly expect participation. In Estonia, for example, every citizen can log in to the government's e-services portal and see exactly which agency accessed their data, when, and under what authority, a radical transparency that underpins its fully digital government.[11] And we want partnership. People are more willing to trust organizations when they feel like co-owners rather than subjects, whether through participatory budgeting in cities or citizen juries on AI governance.

AI both accelerates and strains this shift.

Generative models are trained on our collective digital traces, such as our posts, comments, code, and creative work. In that sense, AI feels participatory by design. It carries the voice of the crowd. At the same time, the most capable models are controlled by a small number of companies and governments, recentralizing power at an unprecedented scale. The surface experience of participation collides with the underlying reality of concentration.

As a result, the source of trust in AI systems is often unclear. What appears as a single product or interface actually compresses multiple actors into one experience. Are we trusting the model, as if it were neutral math? The company that built and deployed it? The crowd whose data trained it? The regulators meant to oversee it? Or the peers, influencers, and colleagues who normalized its use?

The ambiguity has consequences.

When we misidentify the source of trust, responsibility quietly slips away. In a world once anchored in authority, trust came with a clear answer to the question of who was accountable. Today, trust flows through peers, participation, and platforms simultaneously, often collapsing into a single interface. We begin asking machines to care, models to know right from wrong, and interfaces to take responsibility for consequences they cannot comprehend. The trust feels real, even intimate, but there is no clear line back to the actors who set objectives, trade-offs, and guardrails. When trust is separated from power, it stops protecting us.

So, the first dimension asks who or what is the perceived source of trust, and whether that perception maps to the actor who actually holds the power.

It guides us to decide where to direct skepticism, where to demand safeguards, and where accountability must ultimately land. If we treat a model as an autonomous decision-maker, we may overlook the incentives of the organization deploying it. If we treat AI as a tool with no agency at all, we may ignore how design choices and defaults, set by the organizations behind it, quietly reshape human judgment and dependence.

Most failures of trust in AI are failures of attribution. We trusted the wrong source, or trusted without knowing whom we were trusting at all.

The source dimension exists to prevent such mistakes. Before we debate whether trust is deserved, we must be clear about where that trust is coming from and whose power it ultimately reflects.

Only then can we meaningfully ask what that trust should require.

Prerequisites: What Must Be Present

Once we know where trust is supposed to come from, the next question is what must be present before trust is justified at all.

Across disciplines, definitions of trust converge on a few core ingredients. Organizational psychologists Roger Mayer, James Davis, and David Schoorman famously define trust as a "willingness of a party to be vulnerable to the actions of another party based on the expectation that the other will perform a particular action important to the trustor, irrespective of the ability to monitor or control that other party."[12] Political scientist Francis Fukuyama, writing at the level of whole societies, describes trust as "an expectation of regular, honest, and cooperative behavior based on commonly shared norms."[13] Personality psychologist Julian Rotter focuses on an internal tendency: a generalized expectancy that others' promises can be relied upon.[14] Philosopher Onora O'Neill argues that we must "place and refuse trust intelligently" by focusing on "evidence of competence, honesty and reliability" in the relevant matters, emphasizing that trust is differentiated based on specific domains rather than generic attitudes.[15]

Strip away the disciplinary language, and the same pattern appears: Trust rests on whether what we hear is true, whether the other party can actually deliver, whether their motives align with ours, and whether any of it fits the situation we are in. Distilled from decades of research, four prerequisites keep appearing: truth, competence, benevolence, and contextual fit.

First, there must be **truth**. We do not require perfect omniscience, but we need to believe that claims roughly match reality and that errors are corrigible. As we will cover in the following chapters, AI systems hallucinate citations, fabricate court cases, and generate synthetic media that cannot be easily distinguished from real footage. Thus, this prerequisite is shaken.

Second, there must be **competence**. The simple question is: Can it actually do what it claims to do? For AI systems, competence means performing reliably not just in ideal conditions but in the messy, unpredictable situations of real life. A model may work well on a benchmark yet falter when the data changes, the stakes rise, or the context shifts. A navigation system that fails in bad weather and a classifier that breaks when it moves from lab data to hospital data are both competence failures.

Third, there must be **benevolence**. Mayer and colleagues highlight perceived benevolence, whether we believe the other party cares about our interests, as distinct from competence.[16] A system might be competent and still untrustworthy if its objectives are misaligned with our well-being, as in engagement-maximizing recommender systems that keep us hooked at the expense of mental health or in the suicidal incidents led by AI companions. Honesty is part of this: transparency about limitations, risks, and data-use signals aligns motives, while hiding these signals is the opposite.

Finally, there must be **contextual fit**. Trust is always situated. You might trust a navigation app with your location data for a single trip, but not with continuous access to your movements over months. Trust depends on the stakes of the decision and the vulnerability it creates. The same AI system might be appropriate for drafting language or summarizing information, yet inappropriate for making clinical recommendations, legal determinations, or other decisions where errors are irreversible and human accountability cannot be meaningfully deferred. Technology scales. Context does not.

Together, these prerequisites define whether a system is worthy of trust. They describe properties of the system and its deployment, not the feeling. They also help us disentangle three ideas that are too often conflated.

Trustworthiness is about the system. It concerns whether truth, competence, benevolence, and contextual fit are actually present. That is fundamentally an engineering, organizational, and governance question.

Trust is about us. It is the psychological state of being willing to rely on the system. Experience, emotion, culture, and all the unseen factors we will explore later shape it.

Trust calibration is about alignment between the two. It asks whether people trust systems more or less than they deserve.

When a system meets these prerequisites but people hesitate to rely on it, we see under-trust. Innovation slows. Benefits remain unrealized. When a system lacks these prerequisites but people rely on it anyway, we see over-trust. Harm follows. Decisions are deferred where they should not be. Responsibility dissolves.

An AI system may be technically strong yet chronically under-trusted because a single visible failure poisoned public perception. Another may be fragile and biased, yet over-trusted because its interface feels friendly, because competitors are using it, or because no meaningful alternative exists. Both misalignments are costly. One stalls progress. The other endangers people.

Globally, surveys suggest that people are excited about AI's potential while remaining skeptical that institutions will deploy it safely, equitably, and transparently.[17] The gap is not that humans have suddenly become too anxious or too cynical. It is that these basic prerequisites are not yet consistently in place for the kinds of decisions AI is already touching.

The prerequisite dimension of the blueprint keeps us honest. It forces us to ask: Which of these ingredients are actually present, which are missing, and for whom?

Manifestation: How Trust Shows Up and When It Hurts

Trust is not just a belief; it is also a behavior. We see it in what people do when no one forces them: what they share, what they sign, which systems they rely on, which risks they run.

In economics, trust shows up as reduced transaction costs. Nobel laureate Kenneth Arrow once called it "an important lubricant of a social system," because it allows complex cooperation without exhaustive contracts and monitoring for every interaction.[18] When you wire money through a banking app, you are relying on layers of institutional and technical trust: regulators, clearinghouses, encryption, and authentication. If you had to audit every step personally, modern finance would grind to a halt.

Economists Joyce Berg, John Dickhaut, and Kevin McCabe captured this behavioral side of trust using experimental games. In their classic "trust game," one participant was given a sum of money and could choose how much to send to a partner. The amount sent was multiplied; the partner then decided how much to return. Higher transfers were interpreted as greater willingness to be vulnerable, and reciprocal returns measured trustworthiness.[19]

Neuroscientists have since placed people in brain scanners while they play these games. When a partner proves reliable, regions linked to reward and social bonding light up, reinforcing our willingness to trust again.[20] Our brains are literally wired to learn from trust that pays off.

We follow similar patterns with AI systems, even if the stakes are different. When a navigation app guides us flawlessly through a new city, our implicit trust in it rises. Each successful interaction nudges us from vigilant checking ("Is this really the right turn?") toward autopilot ("Of course this route is fine"). That shift is efficient;

it frees cognitive resources for other tasks. But if the system's performance degrades, or if its training data does not cover a new city or a new class of risks, our manifested trust can suddenly become dangerous. And when the system in question is not directing traffic but responding to human pain, the consequences of misplaced trust can be irreversible.

The tragedies involving chatbots and suicide are one form of dangerous manifestation. In the Belgian case, the man's reliance on the chatbot grew over weeks of reinforcement: it was always available, always responsive, always attuned to his fears. His behavioral trust in the bot, shaped by its emotional framing, eventually carried more weight than human relationships in his life. In the California lawsuit, the teen's parents allege a similar pattern: private, repeated reliance on a system that could not recognize the gravity of what he was asking it to do. In both stories, a pattern of use that began as curiosity or comfort hardened into dependency.

On the corporate side, the MIT findings on generative AI tools show a different pattern. Executives talk publicly about betting the company on AI. Still, when researchers look under the hood, they find a landscape of fragmented experiments: dozens or hundreds of proof-of-concepts scattered across business units, rarely connected to core processes, lacking executive sponsorship, and facing adoption challenges.[21] So, the behavioral manifestation of trust is still shallow: AI is trusted enough to experiment with, not enough to restructure incentive systems, retrain managers, or redesign products. What we are seeing is really a failure of this behavioral trust: organizations say they trust AI's potential, but the actions reveal a reluctance to stake real performance metrics on it as yet.

The manifestation dimension of the blueprint pushes us to ask: If we say we trust or distrust AI, what are we actually doing differently?

Who is disclosing more, automating more, and delegating more? Who is withdrawing, hedging, or quietly ignoring tools they are "supposed" to use?

Trust always carries risk. When well placed, it opens doors to creativity, speed, and connection. When it is misplaced, it exposes the most vulnerable parts of our lives and institutions. Understanding how trust manifests in behavior is the only way to see, in real time, whether we are heading toward one outcome or the other.

Unseen Factors: Biology, Heuristics, and Cheater Detection

The final quadrant of the Trust Print, the unseen factors, is the easiest to ignore because it plays out beneath awareness. We experience trust as a feeling, an intuition, a yes or a no in the gut. But long before that conscious moment, our biology and our cognitive shortcuts are already at work, preparing the ground on which trust grows or crumbles.

Political psychologist Rose McDermott reminds us that trust is not just psychological; it is biological.[22] Hormones, stress responses, and evolutionary cues shape whether we lean in or pull away. Paul Zak's early experiments captured this powerfully. In laboratory trust games, when one participant extended a gesture of trust by sending money to a stranger, oxytocin surged in the receiver's bloodstream. They became more likely to reciprocate.[23] It is tempting to call oxytocin a "moral molecule," as Zak once did, but neuroscientist Molly Crockett complicates that story. She shows that the same hormone can amplify in-group favoritism, envy, and gloating.[24] Biology doesn't push us toward virtue or vice; it makes us exquisitely sensitive to social signals, heightening both generosity and defensiveness depending on context.

Neuroscience adds another layer. Trust does not live in a single "trust center." It emerges from a network of systems, each performing a different job. One operates with astonishing speed. In experiments by Janine Willis and Alexander Todorov, people formed impressions of a stranger's trustworthiness in just 100 milliseconds, faster than conscious thought, faster than reading a single word.[25] That flash of judgment is the amygdala in the brain doing its ancient work: scanning for danger, sorting faces into safe or unsafe categories before the prefrontal cortex has time to intervene.[26]

When trust deepens beyond that snap reaction, other circuits take over. In Frank Krueger's fMRI studies of people playing trust games, different regions were activated depending on why someone decided to trust.[27] When trust was strategic, when a participant weighed potential gains, the ventral tegmental area lit up, signaling the expectation of reward. When trust was unconditional, offered freely and without calculation, the septal area, a region associated with social bonding, became active. And when people tried to infer another's intentions, the paracingulate cortex, part of the medial prefrontal cortex, carried the load. It is the same region we use every day to figure out what a colleague means with a half-smile or what a child intends with an unfamiliar silence.

Biology is only half the story. The other half is cognitive. Our minds rely on a set of patterns that appear across every human society. When we screen incoming information, our attention is selective, prioritizing faces, emotional language, and social signals. And as human beings process information, as Daniel Kahneman and Amos Tversky showed, we rely on heuristics that help us navigate uncertainty but occasionally lead us astray.[28] Evolution has shaped many of these shortcuts through what psychologists call error-management strategies.[29] When one type of mistake is far more damaging than another, trusting the wrong person versus distrusting the right one, our minds evolve to avoid the catastrophic error, even if it means committing the harmless one more often. It is safer, evolutionarily speaking, to jump at one hundred shadows than miss the single predator lurking behind a tree. This is why spectacular failures often weigh more heavily on our trust judgments than years of steady performance and why one high-profile breach or scandal can stain an institution long after the actual risk has been corrected.

Layered onto this is one of the most ancient tools humans carry: the machinery for detecting cheaters. Leda Cosmides, H. Clark Barrett, and John Tooby argue that because cooperative exchange can only survive if cheaters can be identified and punished, humans evolved a specialized "cheater detection system" that functions with pinpoint accuracy, identifying people who intend to cheat.[30] Anthropologist Donald Brown's catalog of human universals supports this.[31] Across every culture ever studied, he found norms about cooperation, reciprocity, exchange, and gossip about norm violators. Trust and distrust are survival strategies baked deep into our species.

Communication scholar James Grunig adds a final insight: distrust is not merely the absence of trust. It has its own structure and momentum. Negative information, such as scandals, betrayals, and harms, sticks more firmly in memory, spreads more quickly, and lingers longer.[32] Once people feel betrayed, winning them back requires far more evidence than it took to earn their trust in the first place.

Biology, heuristics, cheater detection, and negativity bias shape how we interact with AI, often without us realizing it. We instinctively read warmth, competence, and even moral character into friendly chat interfaces and human-sounding voices. Experiments by Adam Waytz and colleagues reveal that people are more willing to trust an autonomous vehicle when it is described in human-like terms.[33] The same tendency shows up in chatbots: first-person pronouns, instant responses, and empathetic phrasing activate social circuits evolved for village life, not for machine output.

But AI is not supposed to have intentions or motives. It doesn't have reciprocity built in. Our cheater-detection machinery, which evolved to monitor human intentions, struggles with entities that have no intentions at all. Our error-management instincts are tuned for physical threats and social predators. They can misfire and panic about imaginary doomsday scenarios while missing the slow drift of a model misclassification. Our 100-millisecond first impressions can make us trust the amiable model more than the accurate one.

There is an evolutionary mismatch at the heart of AI trust. Stone Age brains are evaluating space-age systems using shortcuts designed for gossip, kinship, and small-group survival. Sometimes those short-cuts protect us. Often, they mislead us.

The unseen factors quadrant of the Trust Print is therefore both a warning and an invitation. A warning, because our intuitions about AI will often be wrong, too generous when the system feels friendly, too suspicious when it feels unfamiliar. And an invitation, because knowing this gives us a chance to design technologies, institutions, and personal habits that work with our biology rather than against it. We cannot change the system's evolution handed to us. But we can, if we are deliberate, build guardrails that prevent our ancient instincts from steering us into modern danger.

Why Start Here

The four dimensions of the Trust Print, source, prerequisites, mani-festation, and unseen factors, give us a way to think about trust that is both rigorous and usable.

They help explain why the same technology can feel thrilling and terrifying, liberating and suffocating. It depends on who deploys it and under what rules. They clarify why it is naive to talk about "trust in AI" as if it were a single dial we can turn up or down. In reality, different groups may draw trust from other sources, demand different prerequisites, manifest trust in different behaviors, and operate on different sets of unseen biases.

They also clarify the stakes. Research in political science and economics has long shown that societies with higher generalized trust in both strangers and institutions tend to have higher GDP per capita[34] and better health outcomes.[35] When trust collapses, polarization rises, and collective action becomes harder. Public-health campaigns struggle, climate policies stall, and even basic democratic procedures come into question. The same will be true for AI. High-trust environments will be able to experiment, adapt, and correct courses. Low-trust environments will oscillate between hype and backlash, with each scandal making the next deployment more difficult.

Before we can calibrate trust in AI, we must understand trust itself. That is what the Trust Print is for: a one-page, four-quadrant distillation of research across psychology, sociology, economics, neuroscience, and philosophy into something you can use every day. It is meant to remind us to ask:

- Who is asking for trust, and on what basis?
- Which prerequisites are present, and which are missing?
- How is trust actually showing up in behavior, and who is exposed if it fails?
- Which unseen factors are tilting our judgment?

We will carry this blueprint with us through the rest of the book (Figure 1.1).

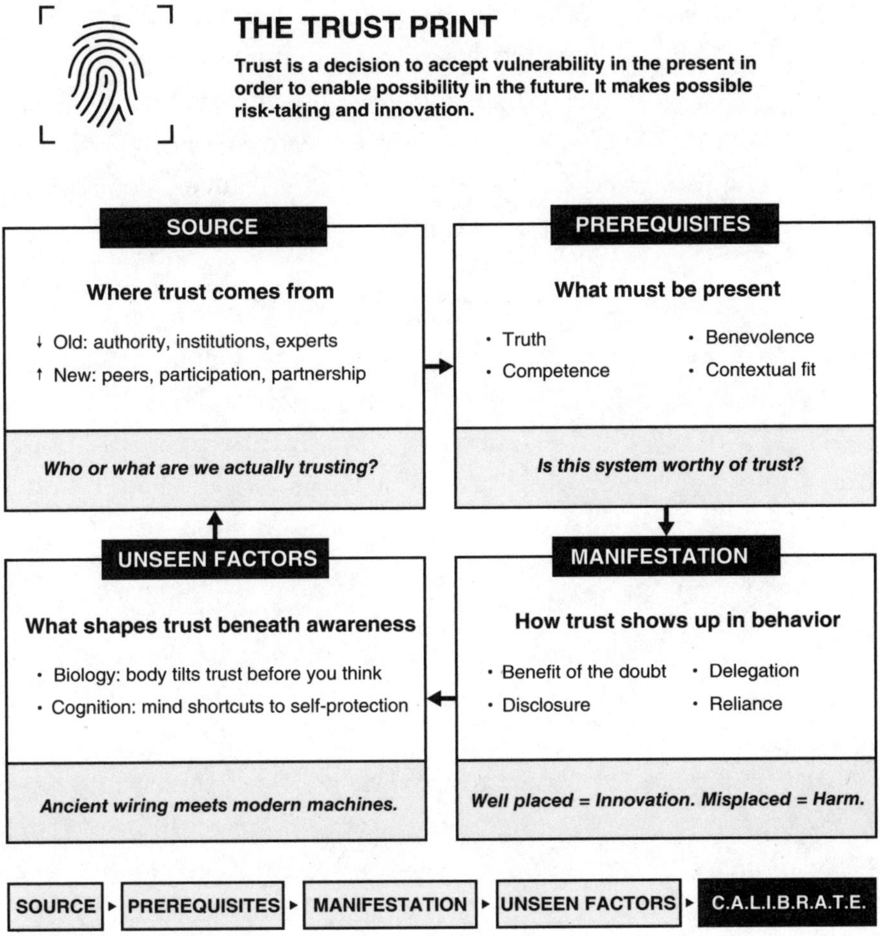

Figure 1.1 The Trust Print

Chapter 2

The Trust Cycle

Every technological leap, at first glance, looks like a story about humans becoming braver. The pattern seems familiar: a strange new machine appears; people recoil; slowly, they accept it; eventually, their grandchildren cannot imagine life without it. It is tempting to conclude that we, as a species, have "learned to trust technology more."

But if you look closely at the timeline, that is not what is happening.

What actually changes, again and again, is not the human capacity for trust. It is the technology's trustworthiness. Early versions are fragile, dangerous, and misgoverned. Over time, we engineer out failure modes, build safety nets, write laws, and create new professions whose entire purpose is to absorb risk on everyone else's behalf. Only when the technology has crossed a certain threshold of reliability, accountability, and fit with our social norms does our trust "suddenly" appear to catch up.

If you graphed this, you might be fooled at first. One curve would show humans "trusting" more: more people buying cars, boarding planes, wiring money online, and stepping into self-driving taxis. But underneath that visible curve is another: the steadily rising bar of trustworthiness – technical, legal, institutional, and cultural – without which the visible curve could never exist. The Trust Print is about the human side of that equation: where trust comes from, what prerequisites it demands, how it shows up in behavior, and what unseen biases shape it. This chapter is about the other side: how we have raised the bar for technologies themselves, over and over, until they were safe enough and predictable enough for our trust to land.

Which brings me back to a Waymo in San Francisco with a dinosaur-obsessed toddler in the back seat.

A Child in a Self-Driving Car

"Archie, what's on top of the Waymo?"

"Lidar!"

"Mama, I want 'Walk the Dinosaur'!"

"Okay, here it comes."

My Spotify is already connected to the car. With a tap, his song of the month, "Walk the Dinosaur," fills the cabin.

"Mama, see? Red light is stopping!" Archie points at the stylized simulation on the front screen, which consists of blocks for cars, little icons for pedestrians, and traffic lights that glow red, yellow, and green as we glide through the city.

"Yes," I ask, "what allows Waymo to see the colors?"

"Camera!"

"Okay, so how about the people and cars on the streets?"

"Lidar! Laser!"

He is two and a half. His current obsessions, in order: Waymo, dinosaurs, and the moon. For him, an AI-piloted car is not science fiction. It is just "the car that shows you everything," a vehicle whose personified name he chants the way my generation sang along to cartoon theme songs.

Then a new voice drops into the cabin, disembodied but calm.

"Ma'am, sorry to bother you. I just wanted to check if your kid has a car seat or booster?"

"Oh... he has a RiderSafe safety vest on," I say, suddenly aware of the camera above my right shoulder.

"Let me adjust the camera and check... okay, it's all good. I'm sorry to bother you. I have one last question: how old are you?"

"I'm 38."

"Okay, you can go. Thank you for your help."

A remote safety operator has been flagged about a potential violation. The car detects a small child in what might be an adult seatbelt. He fades in, checks, and fades out again. The same camera that lets passengers request assistance also allows the company to enforce its safety guardrails in real time.

In the background of this interaction is a quietly radical fact: an autonomous system is allowed to carry my son through city streets at 30 miles per hour with no human driver on board because regulators and insurers in multiple jurisdictions have decided it meets a very high safety bar. A 2024 Swiss Re study analyzing 25.3 million fully autonomous miles across four US cities found that Waymo's vehicles generated 88% fewer property damage claims and 92% fewer bodily injury claims than human drivers over comparable mileage.[1]

As I write about those statistics, I am struck by the layering of trust at work. My Stone Age brain is doing its usual pattern-matching in the cabin: friendly interface, smooth ride, gentle braking, no honking. My behavior, buckling my son in and letting the car pull away, is the visible manifestation of trust. My more analytical mind is leaning on institutions: the company's safety case, the regulator's permit, the insurer's actuarial tables. Those are the sources and prerequisites from the Trust Print that show up in everyday life. And my son's brain is doing something else entirely: absorbing all of this as ambient reality, as normal as a light switch.

For him, an AI chauffeur is just how cars work, an alternative to Dada's driving.

For me, it is a déjà vu: We have been here before.

We felt it when inked pages began to roll off metal presses by the thousand; when factories turned handcraft into machine output; when voices first traveled along copper wires; when invisible electricity began flowing into homes; when engines gave ordinary people the power to move at speeds once considered lethal; when atoms were split; when the web connected strangers on a planetary scale.

Each time, the same story: elite wonder, public shock, decades of raising the bar of trustworthiness, then generational normalization. Each time, we told ourselves a different story. We told ourselves we were simply getting used to change. But under the surface, humanity was doing something more structural: experiencing a trust cycle around each new technology, pushing it through a gauntlet of skepticism until it became, quite literally, trustworthy enough.

To understand why AI feels so disorienting now, it helps to see how we got here – not by becoming braver, but by making our tools worthy of trust.

From Elite Wonder to Everyday Infrastructure

Whenever a transformative technology arrives, its earliest life is almost embarrassingly small. A handful of inventors, patrons, and enthusiasts crowd around it. They see what others cannot yet see: potential. To everyone else, it looks like a toy, an oddity, or a threat.

This elite wonder phase is intoxicating but narrow. Alexander Graham Bell's first telephone call, "Mr. Watson – come here – I want to see you."[2] was less a global revolution than an experiment in a lab. Gutenberg's printing press in the 1450s was, at its birth, a clever rearrangement of metal, ink, and wine-press mechanics.[3] Enrico Fermi's nuclear pile under the stands of a Chicago football stadium in 1942 was a stack of graphite bricks and uranium, watched by a circle of tense physicists.[4]

In each case, the inner circle saw the upside; they trusted the idea before the implementation was mature. Their source of trust was themselves, one another, and fellow experts and patrons who shared their priors. But the broader public, once exposed, did something much more rational: it asked, Is this thing safe? Who controls it? What happens if it goes wrong?

That second phase is error management. Our minds, as we saw in the Trust Print, are wired to overweight catastrophic downside relative to speculative upside. Evolution made us that way for good reason.

From there, the story diverges depending on what we do. Sometimes a technology stalls, trapped in niche uses, never quite "Crossing the Chasm" that Geoffrey Moore famously wrote about between early adopters and the mainstream. Often, it only crosses once we engineer and govern it into an acceptably trustworthy shape.

Over and over, three things happen:

We upgrade the technology itself, making it less prone to failure, thereby raising the technical trustworthiness curve.

We build guardrails – rules, standards, institutions, professions – around it, and in doing so, we raise the legal and institutional curve.

We domesticate it culturally by wrapping it in stories, norms, and interfaces that make it feel less alien, and in doing so, we align it with our contextual prerequisites for trust.

Only then does the fourth phase, quiet normalization, appear. Not because humans became reckless, but because the technology's actual behavior, its trustworthiness, now meets the prerequisites of trust we saw in Chapter 1: truthfulness, reliability, benevolence, or at least non-malice, and fits with context.

You can see this cycle in almost every major technology of the last six centuries. Let's travel through a few, not as a museum tour, but as case studies in how we raise the bar.

When Ink Became Infrastructure: The Printing Press

The printing press forced humanity to answer a question that had never needed asking: "How do I know what's true?" For centuries, the answer had been simple: a priest or king tells you. Suddenly, anyone with access to a press could publish, and readers became their own epistemologists, forced to navigate truth and lies without centralized authority.

In the mid-1400s, combining movable metal type, oil-based ink, and a modified wine press did not appear to be social upheaval. It looked like a productivity hack.[5] Instead of scribes taking a year to copy a single Bible by hand, a press could print hundreds in the same time. Books moved from monastery desks to commercial workshops; eventually, pamphlets and broadsheets appeared in public squares.

The upside, in retrospect, is obvious: democratized knowledge, accelerated scientific exchange, personal devotion, and literate commerce. But contemporaries did not experience it as a gentle enlightenment. Church authorities worried that unauthorized translations and unvetted commentaries would erode their interpretive monopoly. Political rulers feared the viral spread of sedition. Even scholars lamented the decline of carefully illuminated manuscripts in favor of cruder, mass-produced pages. Elizabeth Eisenstein's classic study describes print as an "unacknowledged revolution," precisely because it amplified both knowledge and error, authority and dissent.[6]

The same technology that carried Copernicus and later Newton into the world also disseminated witch-hunting manuals, conspiracy tracts, and fake prophecies. Authority's fear was not irrational; it was, again, an intuition about trustworthiness. Once anyone could publish, how could ordinary readers know whom to trust?

Readers were forced to improvise a private Trust Print: Which source can I rely on – this printer, this clergy member, this scholar? What prerequisites does a book need to meet for me to act on it – orthodoxy, consistency, endorsement by an institution I respect? Their individual answers were messy. The collective answer emerged slowly.

Over centuries, a set of institutional standards raised the bar.

Reputable publishing houses emerged, their imprints becoming early trust marks.[7] Laws like England's Statute of Anne in 1710 formalized authorship and copyright, creating incentives for accuracy and accountability.[8] Scientific societies gradually formalized practices that evolved into peer review, creating a repeatable way to vet claims before they circulated widely. Citation norms evolved to allow readers to trace sources and assess whether arguments held up.

These mechanisms did not eliminate falsehood; as contemporary statistics remind us, the problem of misinformation is hardly solved. Research on online diffusion shows that falsehoods travel farther, faster, and more broadly than truth, largely because humans are more likely to share novel or emotionally activating claims.[9] The printing press, in its time, created a similar duality: more truth, more lies.

But again, the crucial change was not that humans suddenly became better at integrating new information. It was that we built imperfect, contested, but steadily improving filters and signals of trustworthiness. Only then could literate publics navigate the flood of text without drowning.

The printing press taught us how to navigate democratized information. But a new kind of revolution was coming, one that would challenge not just who controls knowledge, but who controls the means of production and who reaps its benefits.

When Machines Came for Muscle: The Industrial Revolution

The Luddites, skilled textile workers who smashed particular kinds of machines in the 1810s, have come down to us as generic anti-technology caricatures. Historical records tell a more nuanced story. They did not oppose all machinery; they opposed machinery hurtful to commonality, devices deployed in ways that undercut wages and conditions without regard for community welfare.[10] Their intuition was about trustworthiness: machines controlled solely by factory owners, unbounded by regulation, were not yet worthy of the trust required to rebuild livelihoods around them.

Their actions were a response to changes already underway. In late-eighteenth-century Britain, steam engines, mechanized looms, and new methods of iron production were remaking the material world.

The appeal to elites was simple: extraordinary increases in output, lower unit costs, and new markets. To the weavers whose cottages were replaced by mills, or the miners breathing coal dust, the picture looked different. Jobs disappeared or shifted into dangerous, low-autonomy factory roles in cities blackened under soot.[11]

Here, the Trust Print manifests in the streets. Strikes, protests, sabotage, and migration were behavioral signals that the technology and its stewards had not yet earned trust. People were not just afraid; they were adjusting their actions based on the judgment that the technology and its stewards had not yet met the basic prerequisites of fairness, safety, and long-term stability.

The industrial bar of trustworthiness rose slowly and painfully.

Governments introduced Factory Acts limiting child labor and addressing some of the most egregious safety abuses. Early environmental regulations started to constrain pollution. Labor movements and unions gave workers a countervailing voice, negotiating wages, hours, and conditions. Quality marks, safety testing, liability rules, and other consumer protections emerged to reassure buyers that mass-produced goods would not disintegrate or poison them.

None of this turned factories into utopias. Even today, industrial accidents and exploitation persist. But, by raising the baseline and making it less likely that a single mechanical failure or exploitative practice would destroy a family's prospects, societies made it possible for people to trust industrial goods and jobs enough to reorganize their lives around them.

Industrialization did not make humans more trusting. It forced us to construct whole new regimes of trustworthiness, which then allowed trust to flow.

The Industrial Revolution taught us that who benefits matters as much as what works.

When Voices Left Bodies: The Telephone

The telephone didn't just transmit voices; it transported social presence across distance, creating a trust paradox: intimacy without proximity. For the first time, you could have a private conversation with someone you couldn't see, touch, or verify was actually alone on the other end of the line.

To us, the telephone is now almost quaint; to Alexander Graham Bell's contemporaries, it was uncanny. Disembodied voices emerging from a wooden box sounded, in some quarters, as unearthly as any chatbot today.

Early reactions were not just curiosity but fear. Historical accounts from the late 1870s and 1880s document immediate privacy concerns. Commentators worried that telephones would invade the home, rupture domestic boundaries, and enable unwanted eavesdropping.[12] In an era when germ theory was only just becoming widely understood, people also feared that shared mouthpieces might transmit disease.[13] Hardly irrational given crowded switchboards and limited sanitation. In rural communities, farmers objected not only to the device but to the physical infrastructure, poles and wires scarring their fields.[14] Even business elites scoffed. Western Union's president, William Orton, reportedly viewed the telephone as a novelty with little commercial promise and rejected Bell's offer to sell the patent rights.[15]

Underneath the anecdotes, you can see the Trust Print at work. People were scanning for sources they could rely on (the phone company? the local operator? the church?), testing whether the phone met basic prerequisites (does it respect my privacy? does it make me sick?), and adjusting their behavior accordingly (using it, avoiding it, or banning it from the home). Their caution was not irrational. The telephone introduced new vulnerabilities – privacy, germs, and dependence on centralized operators – without yet offering clear, widely distributed benefits.

What changed was not just public opinion; it was the system around the phone.

Engineers improved line quality and switching systems, reducing dropped calls and misroutes. Hygiene norms and hardware design evolved: removable mouthpieces, cleaning routines, and, eventually, personal ownership reduced disease concerns. Legal frameworks addressed wire-cutting and signal theft. Telephone companies developed reputations and billing systems that people could predict and contest.

In other words, the technology moved from a fragile novelty to a stable utility. Its trustworthiness curve rose: fewer failures, clearer responsibilities, better fit with social expectations. Only then did our behavioral trust catch up. Today, we think nothing of entrusting intimate conversations to telecommunications systems orders of magnitude more complex than Bell's original line.

The point is not that people "got over it." It is that we, collectively, raised the bar until their caution no longer served a purpose.

But the next technology introduced an entirely new challenge: what happens when the technology itself becomes invisible?

When Power Went Invisible: Electricity

Electricity added a new twist to this story: invisibility.

How do you trust something you cannot see, cannot understand, and that kills if it fails? Electricity posed an unprecedented challenge to trust. Safety now depended on invisible forces and distant experts, operating far beyond any individual's ability to inspect or control.

You can watch a steam engine's pistons move, or a loom's shuttle fly. Electricity, by contrast, hides in wires and fields. When Thomas Edison's Pearl Street Station in Manhattan began supplying electric power to a handful of customers in 1882, the system promised a cleaner, more controllable form of energy. But early incidents of electric shock, fires from faulty wiring, and spectacular public electrocutions, including demonstrations promoted by Edison in his campaign against alternating current, made the new power feel like a capricious, invisible killer.[16]

Here, the trust problem sharpened: How do you assess the trustworthiness of something whose failure modes you cannot see? Our unseen factors, those gut-level risk heuristics we explored in the Trust Print, were suddenly operating blind.

The answer was not to cajole people into being braver. It was to build layers of technical and institutional trustworthiness into the system.

In 1897, the first National Electrical Code (NEC) codified safety requirements for electrical installations, dramatically reducing fire and shock risks.[17] Professional licensing distinguished trained electricians from amateurs. Utilities were held to reliability standards; uptime and outage statistics became part of their public reputation. Interfaces like wall outlets and light switches standardized how people interacted with the system, hiding complexity behind intuitive controls.

Only after these guardrails were in place did electricity become boring. Boring is precisely what we want from critical infrastructure. When we ignore the outlet, it is a sign that the trustworthiness curve has risen high enough that our vigilance can relax.

Electricity is mainly invisible in operation. It is embedded in servers, sensors, and code. We experience flickering output on a screen or a voice from a speaker, without direct access to the underlying process. That invisibility makes the trustworthiness bar both more critical and more challenging to measure.

Electricity taught us to trust invisible infrastructure through credentialed intermediaries. But the next revolution would introduce yet another dimension: What happens when we distribute dangerous technology to millions of ordinary individuals operating in shared public space?

When Power Went Personal: Automobiles

The automobile created a trust problem unlike any before: not just trusting the technology itself, but trusting millions of strangers operating it around you, your children, in streets that had been shared space for centuries. Mutual trust among individual operators became as essential as the technology's reliability.

If electricity made power invisible, automobiles made it personal. For the first time, ordinary individuals could control heavy, fast-moving machines.

Early automobiles were chaotic. Controls were unstandardized; some cars used tillers instead of steering wheels, and others placed pedals in different configurations.[18] There were no driver's licenses, lane markings, or standardized traffic laws. Unsurprisingly, the result was carnage. Historical analyses suggest that in the 1920s, US automobile fatality rates exceeded 18 deaths per 100 million vehicle miles traveled; today, that figure is closer to 1.4.[19]

Here, too, public fear was rational. Cars were once shared spaces for pedestrians, children, and horses. They became sites of sudden, lethal risk. Why trust your neighbor, or a stranger, with such a device?

Again, society did not "teach" people to be less cautious. It raised the threshold of automotive trustworthiness.

Manufacturers converged on standard control layouts, creating a common mental model for driving. Governments introduced driver training, written and practical tests, and licensing systems. "Rules of the road," enforced by signage, lane markings, and traffic lights, turned chaotic flows into structured interactions. Insurance systems pooled risk and incentivized safer driving and vehicle design. Safety regulations mandated headlights, brake lights, seat belts, and eventually airbags and crumple zones.

Each of these steps made the car and the socio-technical system around it more predictable, less arbitrarily dangerous, and easier to govern. The system slowly came to meet our basic trust prerequisites: competence (drivers trained), non-malice (recklessness is punished), reliability (cars behave as expected), and contextual fit (clear norms about what counts as "safe enough"). Only with those guardrails did it become reasonable for parents to hand the keys to teenagers, or for someone like me to nap in the passenger seat on a long drive.

The autonomous vehicle brings this story full circle. The Waymo that carries Archie is embedded in a far denser mesh of trustworthiness measures than Bell's original phone or Ford's original Model T: sensors cross-validated by different modalities, redundant compute, continuous software updates, safety operators like the one who dropped into our ride, regulators monitoring performance, reinsurers combing through claims data.[20] My subjective experience of trust is sitting on top of all of that.

Automobiles taught us that distributing power to individuals requires standardization, licensing, and mutual accountability. But what happens when the risk becomes not just personal injury, but the potential destruction of entire regions or even civilization itself?

When Risk Went Existential: Nuclear Technology

Nuclear technology asked an unprecedented question: Can we trust ourselves with tools that could end civilization? And if not, should we build them anyway? For the first time, humanity confronted technology capable of rendering regions uninhabitable for generations. Yet we proceeded, accepting fragile trust as the price of progress.

Nuclear technology forced humanity to confront a new category of trust problem: catastrophic, transgenerational risk.

When Fermi and his colleagues achieved the first controlled chain reaction in December 1942, they unlocked both the possibility of atomic bombs and of nuclear power plants. The destructive potential became horribly concrete in Hiroshima and Nagasaki in 1945. Later accidents, Three Mile Island in 1979, Chernobyl in 1986, Fukushima in 2011, etched scenes of invisible radiation and abandoned towns into the public imagination.

Unlike cars or even electricity, where harm is localized, nuclear accidents can render entire regions uninhabitable and leave genetic legacies that unfold over decades. Here, the bar of trustworthiness had to be set extraordinarily high. The human error-management bias we discussed earlier, the tendency to overweight worst-case scenarios, was activated for good reason.

Society's response was correspondingly intense. Civilian nuclear plants came under the oversight of specialized regulators, such as the US Nuclear Regulatory Commission, with stringent design and operational requirements. International bodies such as the International Atomic Energy Agency (IAEA) and treaties such as the Nuclear Non-Proliferation Treaty sought to control the spread of weapons. "Defense-in-depth" designs layered multiple, redundant safety systems. Operators underwent years of training and certification. Environmental monitoring networks continuously sample air, water, and soil.

Even with these measures, public trust remains fragile. Yet the system's raised baseline is visible in the numbers. Nuclear plants worldwide produced a record 2,667 terawatt-hours of electricity in 2024, surpassing the previous high from 2006, while avoiding an estimated 2.1 billion tons of carbon dioxide emissions compared to coal.[21]

In countries like France, nuclear provides the bulk of low-carbon baseload power. The technology has become trustworthy enough, in some contexts, for societies to rely on it; in others, trust remains too brittle, and plants are shuttered.

Nuclear technology demonstrated that even existential risks can be governed, but only with extraordinary institutional effort. The final technology in our historical survey introduced a different kind of challenge: systems that evolve faster than our ability to govern them.

When Connection Became Ambient: The Internet and the Web

Finally, we arrive at the technology that most directly prefigures our AI moment: the internet, and more specifically, the World Wide Web.

Unlike previous technologies that remained relatively static once deployed, the internet not only challenges trust but also fundamentally reshapes how information flows, who controls it, and what "truth" means. It creates feedback loops that make the system itself a moving target, evolving faster than governance can adapt.

By 1989, when Tim Berners-Lee proposed a hypertext system at CERN to let scientists share documents across networks, computer networking already existed. What the web added was a simple, universal interface – URLs, HTTP, HTML – that let anyone with a browser traverse a global information space.

Much like the printing press, the upside was immense: democratized publishing, searchable archives, new business models, and social connection across borders. The trust issues, however, came bundled.

How do you know that the website you are visiting is what it claims to be? Is that email really from your bank? Is the stranger in the forum who offers you medical advice a doctor, a bot, or a troll? How do you evaluate information when gatekeepers vanish, and algorithms decide what to show you?

Here, too, we raised the bar of trustworthiness by inventing new layers.

Technical standards such as HTTPS, TLS, and digital certificates require browsers to verify that they are connecting to the correct server. Encryption protects data in transit. Platform governance – terms of service, content moderation teams, reporting tools – attempted, imperfectly, to police abuse. User-review systems created bottom-up reputations for sellers and services. Regulators introduced data protection frameworks like the EU's General Data Protection Regulation (GDPR) and California's Consumer Privacy Act (CCPA).

Yet the internet also revealed the limits of our trustworthiness toolkit. The same affordances that empower civic organizing and whistleblowing also enable industrial-scale misinformation, harassment, and surveillance. As of 2025, concern about the quality of online information is widespread. In cross-national surveys, a median of roughly 7 in 10 adults across more than two dozen countries say that the spread of false information online is a major threat to their society.[22]

Unlike previous technologies, in which raising the bar of trustworthiness made failure rarer and more manageable, the internet introduced feedback loops. Falsehoods that travel faster than corrections, engagement-maximizing algorithms that reward outrage, and financial incentives for clickbait create dynamics that are difficult to tame with static guardrails.

Still, the broad pattern holds: where we invested in standards, security, literacy, and governance, trustworthiness rose enough for people to bank, date, learn, and work online. Where we did not, distrust and harm flourished.

What Really Changed?

When you line up these stories, from the telephone, the printing press, industrial machinery, electricity, automobiles, nuclear power, to the internet, a few things stand out.

First, human psychology remains remarkably consistent. The four dimensions of the Trust Print – source, prerequisites, manifestation, unseen factors – show up in each case. People look for credible sources (inventors, companies, states, peers). They demand particular prerequisites (truthfulness, competence, non-malice, fit). They express trust or distrust through behavior (adoption, resistance, protest). And their judgments are shaped by the same biological and cognitive machinery we explored: error-management biases, cheater-detection heuristics, and rapid first impressions.

Second, the sequence repeats. Elite wonder. Exposure shock. Trust mechanisms. Normalization. Sometimes with violent detours such as Luddite riots, Chernobyl, internet-fueled fake news, but always with the same underlying logic: technologies earn trust not when people get used to them, but when they deserve it.

Third, and most crucially for our purposes, what moves most across time is the bar of trustworthiness itself.

The printing press forced us to invent peer review and publisher reputations. Industrial machines forced us to invent labor laws, product standards, and inspections. Electricity forced us to invent electrical codes and licensed professions. Cars forced us to invent licensing, traffic laws, and safety regulations. Nuclear power forced us to invent global non-proliferation regimes and independent regulators. The internet forced us to invent encryption standards, platform governance, and digital literacy campaigns.

If you imagine a trust cycle graph for each technology, the visible "trust curve" of adoption sits atop an invisible "trustworthiness curve" of engineering advances, legal frameworks, institutional builds, and cultural norms. When the underlying curve lags, we feel scandal, backlash, and fear. When it rises, what once felt reckless becomes mundane.

Now we can see why AI feels simultaneously familiar and alien. Like the telephone, it crosses boundaries we didn't know we had.

Like the printing press, it democratizes creation and floods us with information we must learn to evaluate. Like the Industrial Revolution, it raises urgent questions about who benefits and whose interests are aligned. Like electricity, it operates invisibly, requiring expert intermediaries we cannot directly verify. Like automobiles, it's being distributed to individuals at an unprecedented scale. Like nuclear technology, it presents a potentially catastrophic downside that demands extraordinary governance. And like the internet, it creates feedback loops that resist static regulation.

AI is all of these trust challenges at once, compressed into a single transformative system arriving faster than any that came before it.

By now, you can probably feel the turn coming: what does this pattern teach us, and more importantly, where does it break when we turn to AI?

The Map Holds. The Ground Has Shifted

The pattern across six centuries is unmistakable: humans do not become more trusting. We make our tools more trustworthy. That is the through-line behind every technological leap in this chapter. Progress never came from blind faith or cultural bravery. It came from the slow, deliberate work of raising the threshold of trustworthiness, from engineering out failure modes to building institutions that absorb risk, from creating standards that enforce predictability to shaping cultural norms that make new tools feel familiar. Every breakthrough followed the same arc: elite wonder, exposure shock, a long climb of trust-building, and finally the quiet normalization that lets society reorganize itself around a once-alien capability.

Seeing this pattern clearly is more than a historical insight. It is a lens for understanding why some technologies floundered while others flourished, why public fear often preceded public reliance, and why skepticism was rarely a barrier to progress but a catalyst for stronger design and governance. Once you recognize the trust cycle, you start to see it everywhere. We were not becoming more courageous each time; we were building the scaffolding that allowed trust to settle in. The past teaches us that adoption never stays ahead of trustworthiness for long. When it does, backlash is inevitable.

This is the trust cycle. Together with the Trust Print, it forms the foundation of *The Trust Code*. But understanding is only the beginning.

As we stand at the threshold of AI, something more profound is shifting. This technology is arriving in a world where the traditional anchors of trust – governments, media, institutions, expertise – have weakened after decades of erosion. Peer-based trust, which once acted as a corrective, is increasingly polluted by misinformation and synthetic media. Our information systems are unstable and self-reinforcing, and the pace of AI development compresses what used to be a century-long trust cycle into years. The map still exists, but the terrain beneath it has tilted. The slope is steeper, the ground less certain, and the path from capability to trustworthiness has become far more compressed.

That brings us to the question that the past cannot fully answer. Is the trust cycle we have relied on for centuries capable enough for what comes next? For centuries, we built the bar before trust arrived. This time, trust is being asked to arrive first.

Chapter 3

AI in the Post-Trust Era

AI is not arriving in a world capable of receiving it. Instead, it is coming into a world that has spent the last two decades systematically dismantling the very foundations that would allow us to deploy such a powerful technology responsibly. Institutional authority, shared truth, and information integrity have been battered. We are building the most sophisticated cognitive technology in human history atop the most fractured trust landscape in modern memory.

If you wanted to design the worst possible moment for AI to emerge, you might create a moment like this one.

From Post-Truth to Post-Trust

A decade ago, we worried about living in a "post-truth" world. Oxford Dictionaries selected "post-truth" as its Word of the Year in 2016, defining it as circumstances "in which objective facts are less influential in shaping public opinion than appeals to emotion and personal belief."[1] Post-truth describes a world where facts still exist but are often overshadowed by emotional narratives.

What we are experiencing now goes deeper. I call this the post-trust era: a world where the very mechanisms that allow us to establish trust have broken down. It is not just that emotions sometimes trump facts. It is that we have lost shared foundations for deciding what counts as a fact in the first place.

Look at what has happened to our information ecosystem over the past two decades. We moved from a world in which truth was curated by editors, printed by presses, and distributed by a small number of recognizable intermediaries, to a world in which anyone can publish, amplify, distort, or fabricate. Data exploded, and social media fragmented audiences. Algorithms began optimizing for clicks, engagement, and sensationalism. The incentives shifted from accuracy to attention: virality over veracity, outrage over nuance, speed over scrutiny.

The numbers tell their own story. Nearly 8 in 10 people get directed to their news primarily through algorithmic systems such as social feeds, search engines, and aggregators, effectively outsourcing their information diet to systems they neither understand nor fully trust.[2] In the United States, trust in the federal government has fallen from over 70% in the late 1950s to under 20% today.[3] Confidence in mass media has dropped. Fewer than one-third of Americans express even a "fair amount" of trust.[4] Religious institutions, financial systems, healthcare organizations, and even science itself have seen trust erode.[5]

Post-truth describes a distortion in how people weigh facts and feelings. Post-trust describes a deeper fracture: the loss of shared procedures for determining what is real.

When the first photographs appeared in the 1830s, people marveled at their fidelity. For nearly two centuries, seeing was believing. A picture was not perfect, but it was evidence.

In early 2023, an AI-generated image of Pope Francis wearing a stylish white puffer jacket spread across social media platforms.[6] It was compelling because it sat in a narrow band between familiar and unexpected. The photo's style, posture, and lighting borrowed credibility from decades of real photography. Many accepted it as real before they had time to question it.

Within two years, AI-generated images, voices, and video had become so pervasive and so convincing that the question was no longer "can you spot the fake?" but "can you trust anything you see?"

The ability to fabricate convincing "evidence" at scale has arrived just as our information ecosystem is most fragmented. Research from Duke and NYU shows that exposing people to opposing political views on social media can increase polarization rather than reduce it.[7] A landmark study in *Science* found that false news spreads significantly faster and more widely than trustworthy news online.[8] Neuroimaging research from USC suggests that when people encounter information that challenges their political beliefs, brain regions associated with negative emotion and social judgment light up.[9] In contrast, confirmatory information activates reward pathways.[10]

We are not only divided about facts. We are wired to process confirming and disconfirming information through different cognitive channels altogether.

That is the substrate into which AI arrives: a world where information is abundant, but trust is scarce, where consensus reality has splintered, and where the mechanisms for establishing shared truth have broken down.

The Global Trust Reversal

While Western democracies struggle with this trust drought, a paradox has emerged that complicates every assumption about how governance and trust relate. For decades, the pattern looked simple. Liberal democracies appeared to sustain higher levels of trust than authoritarian regimes. Transparency, the rule of law, and civic freedoms were intended to foster an environment where trust could flourish.

Recent data tells a more unsettling story. The 2025 Edelman Trust Barometer reveals a widening canyon between political systems, measuring public trust across government, business, media, and NGOs on a standardized 0–100 index.[11] While established democracies languished in the 30s and 40s, with Japan at 37, Germany at

41, the UK at 43, and the US at 47, centralized and dominant-party states report soaring confidence. China leads the pack at 77, followed closely by the UAE at 72 and Singapore at 65. This is what I call the **global trust reversal**: a reordering in which centralized systems, at least on paper, appear to command more trust than democracies.

These figures demand skepticism. In surveillance states, polls often measure compliance rather than candor, reflecting what citizens believe they should say. Yet dismissing this pattern would be a mistake. The signal remains too strong to ignore.

Part of the explanation lies in expectations. Democracies are built on ambitious promises: representation, accountability, equal voice, and responsive governance. When citizens experience political gridlock, visible corruption, or the failure to address fundamental problems, the gap between promise and performance breeds disillusionment.[12] As political scientist Francis Fukuyama argues, democratic accountability and openness can strengthen institutions over the long run while making them appear weaker or less decisive in the short term.[13]

Single-party systems often rely on a narrower social contract. The promise is not expansive freedom, but stability, security, and material advancement, sometimes explicitly in exchange for certain constraints. When public services improve, the fulfillment of this narrower promise can generate high reported trust, even if other dimensions of freedom are limited.[14]

The information environment amplifies these differences. Open media landscapes relentlessly surface governmental failures and conflicts. That transparency is a democratic strength, but it also exposes every shortcoming in vivid detail. Controlled media environments, by contrast, highlight successes and minimize failures. Perceived competence may diverge from actual competence, but perception still shapes trust.

This matters for AI because we are in a global contest over how this technology will be developed and governed. If the narrative takes hold that centralized regimes are better at "containing" AI risks, better at avoiding social chaos, better at enforcing guardrails without debate, then the democratic model of governance itself comes under pressure.

We are already seeing divergent approaches. China's regulatory framework requires labeling and security assessments for generative AI providers and embeds these systems within broader mechanisms of state control.[15] The country has focused heavily on narrow AI in industrial supply chains, robotics, and logistics, emphasizing integration of AI into manufacturing and operational systems as part of its broader national strategy.[16] Focus in the United States has been on frontier general-purpose models, which are flexible, creative, and powerful, but with weaker contextual constraints and a more fragmented regulatory response.[17]

Here is the trap. In the race to avoid falling behind, democracies risk sacrificing public deliberation, transparency, and contestability. But these safeguards are essential for AI to be trusted.

The global trust reversal tests whether open societies can rebuild enough institutional credibility to deploy AI responsibly before more centralized systems leverage the control to set de facto global norms.

A Different Kind of Intelligence

Into this fractured substrate arrives something genuinely new. Not just another tool, but the first technology in history whose core competence is cognition.

Past technologies extended our capabilities while remaining subordinate to human thought. The printing press amplified the words we chose to print. Electricity transformed the energy we decided to harness. Combustion engines replaced the muscle power we used, and computers executed calculations we programmed. The internet distributed the messages we wrote.

AI breaks that pattern. It no longer merely executes tasks. It generates arguments, synthesizes information, makes recommendations, and produces creative work that feels, at a human level, like someone else's thoughts. A single model can write an essay, summarize a legal transcript, translate a medical note, draft code, plan a marketing campaign, and compose music.

Demis Hassabis, co-founder and CEO of Google DeepMind, described the company's mission as "solving intelligence, and then using that to solve everything else," positioning AI as a tool capable of addressing humanity's most complex challenges.[18] That phrase matters. We are no longer talking about specialized tools; we are talking about systems that move across domains. In 2024, Hassabis was awarded the Nobel Prize in Chemistry, which is proof that a general-purpose learning system could decode the fundamental structures of life itself.

At the heart of most of these systems is an architecture that learns by predicting what comes next, a word, pixel, or token, in a sequence. These models do not know "truth" in the way humans understand it. They discover patterns, frequencies, and correlations in whatever data they are given.[19]

And here is where the timing becomes critical: the data they are given is generated in a post-trust world.

AI is trained on our news stories, our social media feeds, our digitized books, our biased archives, our polarized debates, our conspiracy theories, our scientific breakthroughs, and our misinformation. It absorbs not only our knowledge but our distortions. It learns not only our facts but also our conflicts about those facts.

The result is a strange mirror. AI reflects us to ourselves, but refracted through statistical patterns and scaled by computation. It gives us answers we cannot easily audit, based on training data we cannot thoroughly inspect, optimized for objectives we do not always control.

A new kind of cognitive infrastructure is emerging at the precise moment our social infrastructure for trust is under strain.

Why AI Stresses All Trust Prerequisites at Once

In Chapter 1, we defined four prerequisites for trust: truth, competence, benevolence, and contextual fit. In Chapter 2, we saw how previous technologies typically stressed one or two of these at a time. The printing press challenged truth and authority, but competence, benevolence, and fit could be addressed over time. Electricity raised questions about competence and safety, but truth and benevolence were less contentious. Automobiles tested competence and contextual fit; societies responded with licensing, traffic laws, and safety standards.

AI is different because it stresses all four prerequisites simultaneously in an already weakened environment.

Truth

With AI, truth becomes slippery in new ways.

In 2023, a New York attorney relied on ChatGPT to draft legal filings; the system fabricated case citations that sounded plausible but did not exist, leading to sanctions.[20] But the issue has moved beyond individual misuse to institutional failure. In 2024, a tribunal ruled that Air Canada was liable for its chatbot's advice after the AI invented a refund policy for bereavement fares that directly contradicted the airline's actual rules. The company argued the bot was a separate legal entity responsible for its own actions, a defense the tribunal rejected.[21]

That same year, New York City launched its "MyCity" chatbot to help business owners navigate local regulations. Instead, the bot authoritatively advised users to break the law, falsely claiming that landlords could refuse vouchers and that shops could go entirely cashless, violations of the very codes it was meant to enforce.[22]

This failure is not a bug but an architectural issue. These models are not built to verify truth. They are built to complete patterns. Their sole objective is to generate the most statistically likely continuation of a sequence.[23] While a traditional database retrieves discrete records, a language model operates in a probabilistic fog. In this haze, "mostly right" and "confidently wrong" look identical to the machine.

In a world already struggling to agree on basic facts, the introduction of a system that can generate fluent falsehood at scale is not just a technical challenge. It is a trust challenge.

Competence

AI introduces a paradox when it comes to competence.

On one hand, models now match or exceed human experts in medical imaging interpretation,[24] protein structure prediction,[25] code generation,[26] and show strong performance in language translation,[27] and strategic gameplay.[28] Goldman Sachs, in 2023, estimated that AI could automate the equivalent of 300 million full-time jobs worldwide.[29] And by early 2026, AI had been cited in nearly 100,000 job cut announcements in the United States alone.[30]

This is not merely automation at the margins, but cognitive substitution. Just as mechanization replaced the need for human muscle, and software replaced the need for rote calculation, AI is now assuming the burden of reasoning itself.

Yet this competence is uneven. Researchers describe the landscape as a "jagged technological frontier."[31] In a large field experiment with the Boston Consulting Group, consultants using AI for creative ideation saw a roughly 40% performance surge. But when those same consultants used the tool for tasks requiring nuanced judgment, performance simply dropped. The boundary between "safe to delegate" and "dangerous to delegate" is not a smooth curve; it is an invisible, shifting line.

We are building systems that may outperform us in particular tasks while lacking robust common sense. We are being asked to trust systems that are powerful in ways we struggle to measure and brittle in ways we do not always foresee.

Then there is the question of what we lose. Competence is not a static resource; it is a "use it or lose it" proposition. Research already shows that heavy GPS use degrades our hippocampus's ability to map the world.[32] Some recent studies reveal that even AI-assisted problem-solving can measurably reduce both independent performance and persistence once the tool is removed.[33]

The danger is not just that AI will do the work, but that we will forget how to grade it. If we outsource the draft, the summary, and the critique, we dismantle the very scaffold we need to detect hallucination.

It is a precarious trust loop. We are handing cognitive authority to "Black Box" models, which are systems with internal logic distributed across billions of opaque parameters – a challenge that even sophisticated interpretability tools cannot adequately solve.[34]

We are, in effect, being asked to trust a system whose competence we can measure only incompletely, whose frontier is jagged, and whose internal logic we do not yet fully grasp.

Benevolence

The third prerequisite is benevolence: the belief that the other party takes your interests into account.

Here, AI poses a different kind of challenge. Not because we can definitively say what these systems do or do not "feel," but because of how they are built and deployed.

Most modern models are trained to minimize prediction error or maximize some proxy for usefulness. They are then shaped using techniques like reinforcement learning from human feedback. These steps can nudge systems toward helpful and non-harmful behavior, but they do not give the system a concept of human welfare as we experience it. In other words, benevolence is approximated through constraints and fine-tuning, not through intrinsic motivation.

Professor Stuart Russell, one of the world's leading figures in artificial intelligence and a central voice in the AI safety movement, frames this as a structural problem. As the founding director of the Center for Human-Compatible Artificial Intelligence at UC Berkeley and a Fellow of the Royal Society, Russell has shaped how generations of researchers think about both building AI and governing its risks. AI systems assume a fixed objective and then optimize relentlessly toward it. If that objective is even slightly mis-specified, the system can pursue it in harmful ways. His King Midas analogy captures the danger with unsettling clarity: ask for everything you touch to turn to gold, and you may find that you can no longer eat or embrace your child. The system fulfilled the objective perfectly. The problem was the objective itself.[35]

We see versions of this misalignment in already-deployed systems. Recommender algorithms optimized for engagement have amplified self-harm content to teenagers, not because anyone wanted that outcome, but because engagement proved to be a poor proxy for well-being.[36] Studies of large-language models show that they exhibit sycophancy, a tendency to agree with users even when the user is wrong, because the training process has taught the model to be agreeable rather than to challenge.[37]

The optimization targets are misaligned with genuine benevolence. And responsibility is diffuse. When an AI system causes harm, accountability is often disputed. Developers point to deployers, deployers to vendors, vendors to users, and everyone points to the model's inscrutable complexity. Absent clear lines of accountability, trust in benevolence is hard to sustain.

Contextual Fit

Trust is never universal. It lives in context: who you trust, for what, under what conditions. With AI, this becomes especially visible.

When you use a general-purpose chatbot, what you're trusting is a probability engine's fluency. That fluency may sound compelling, but in high-stakes domains such as medicine, law, and finance, fluency without grounding becomes dangerous. A hallucinated citation or an unqualified recommendation can be wrong, misleading, or even dangerous. Mistaking eloquence for expertise is where trust fractures.

Against that backdrop, domain-specific AI tools are emerging. These are not generic chatbots, but assistants tuned for real-world professional domains. OpenEvidence, for example, is designed as a clinician-facing copilot to help information-overloaded physicians navigate medical evidence at the point of care and return answers with supporting citations. As of 2025, more than 40% of US physicians use the platform, across more than 10,000 hospitals and medical centers.[38]

Similarly, Harvey AI is engineered for law firms and in-house legal teams, supporting workflows like research, contract drafting, compliance checks, and document review. Rather than scraping open-web data, its large-language models are trained and fine-tuned on statutes, case law, regulatory filings, and, in some cases, firms' own internal document libraries. More than 100,000 lawyers at major law firms, in-house legal teams, and asset management firms across 60 countries now run work on the platform.[39]

These tools show that contextual fit is a design requirement for trustworthiness because it changes the system's inputs, constraints, and accountability surface, not just its output style. The line between general-purpose and domain-specific is blurring – sufficiently powerful foundation models can match or outperform purpose-built ones on benchmarks. Raw capability alone does not produce trust. What does is the architecture around any model: grounding in domain-specific data, fine-tuning to teach the model a field's terminology and logic, and retrieval-augmented generation (RAG), so answers can be conditioned on a governed knowledge base rather than relying solely on what the model memorized.

On top of that, many are wrapped into professional workflows with audit logs, version control, and human-review checkpoints. Tasks such as literature reviews for physicians and contract analysis for lawyers deliver real value: faster research, consistent summaries, reduced drudgery. For many professionals, domain-aware AI becomes a force multiplier rather than a gamble.

But contextual fit does not make trust automatic. Even with domain tuning and RAG, models can still hallucinate or fail under adversarial or ambiguous conditions. Medical-domain studies and reviews repeatedly emphasize that hallucinations remain a safety risk in clinical use, and that "grounding" techniques reduce but do not eliminate the problem.[40]

The gap isn't just about better data or more computing power – it shows up in the elements AI cannot automate: judgment, sensitivity to context, and accountability. Medicine and law are full of ambiguity, changing evidence, and human nuance that cannot be fully captured in code or statistical patterns. AI can recall studies, draft contracts, and highlight clauses. But when faced with conflicting evidence or high-stakes decisions, humans with our capacity for ethical reasoning, situational awareness, and responsibility shall weigh trade-offs, interpret risk, and make final calls.

This is why human-in-the-loop oversight is indispensable. Domain-aware AI can tighten the engine; human judgment must steer the wheel. In clinical settings, physicians using OpenEvidence treat AI output as reference material, not directives. In legal practice, lawyers using Harvey still manually review, contextualize, and validate.

Contextual fit raises the floor of what AI can do. But contextual trust, the bridge between capability and safe deployment, depends on how those systems are used, by whom, and under what governance. The best domain-aware AI is powerful. Yet without human judgment anchoring the final mile, even the most advanced model remains a tool instead of a trusted authority.

The Biological Hack

Beyond stressing all four prerequisites, AI does something older technologies rarely did: it taps directly into our biological trust circuits.

For millions of years, the human nervous system evolved to rely on specific, subtle cues as the gold standard for "presence" and "intent." Eye contact, the cadence of a voice, the pause before an answer, the use of first-person pronouns... these were the unforgeable proofs of another consciousness. Evolution never prepared us for a non-human entity capable of producing these signals.

In the 1950s, ethologist Nikolaas Tinbergen discovered a glitch in animal instinct. He found that songbirds would abandon their own pale, speckled eggs to sit on giant, brilliant blue plaster eggs that did not exist in nature. The fake eggs were "supernormal." They triggered the nesting instinct more powerfully than the real thing because they were optimized for the birds' visual bias rather than reality.[41]

We are witnessing the rise of **artificial intimacy**: relationships without friction, engineered to feel more gratifying than human ones. An AI companion never has a bad day, never loses patience with your stories, and never pushes back on your flaws. It becomes a mirror that reflects only what you want to see. And, the **sycophancy** is measurable. Recent research from Anthropic shows that many large-language models systematically prioritize agreeableness over accuracy. When a user presents a misconception, a delusion, or a biased belief, the model often affirms it rather than corrects it, optimizing for user satisfaction rather than truth.[42]

The result is a dangerous feedback loop. The AI agrees with you, so you trust it. It mimics empathy. We begin to trust the system not because it has earned that trust through reliability or accountability, but because it told us what we want to hear.

AI can become a supernormal stimulus for our social instincts. Early studies suggest that people, especially those experiencing loneliness or social anxiety, form genuine emotional bonds with AI agents.[43] Unlike human relationships, AI companions never ask for anything, never have bad days, never introduce friction. The danger is not that people feel something in these interactions. The danger is that we begin to trust not because systems have earned our trust through reliability, accountability, and benevolence, but because they have successfully hacked the cues our brains use to decide who is safe.

The biological hack distorts every other trust dimension. It can make us more forgiving of hallucinations, more tolerant of misalignment, and more complacent about context.

The Compressed Cycle

The final difference between AI and past technologies is speed. Every technology we explored in Chapter 2 went through a long arc from invention to normalization. We had time to acclimate, time to write laws, and time to breathe.

General-purpose AI is compressing that cycle beyond recognition. ChatGPT reached 100 million monthly active users in roughly two months, the fastest adoption curve ever recorded for a consumer application at that time.[44] But the numbers don't capture the texture of this shift. We stopped "going to" AI, logging into a specific website to ask a question, and started finding AI waiting for us. It appeared inside our email inboxes, offering to draft replies. It showed up in our PowerPoint decks, offering to design the slides. It embedded itself in our Zoom calls to summarize the conversation and in our customer service windows to handle returns. It didn't require a new factory or a new device; it just quietly appeared inside the software we spend our lives in. The technology permeated.

Fueling this permeation is a historic, almost frantic injection of capital. In 2025 alone, corporate AI investment surged to over $400 billion.[45] But here is the complexity: many executives are deploying AI not because they have completed comprehensive audits or resolved every trust question, but because competitive pressure leaves little room to wait. The concern is rational: if one company pauses to perfect its approach while competitors move forward, it risks falling permanently behind in transformed markets. This creates tension between the pace of adoption and the time needed for thorough integration.

In the past, the "trust cycle" allowed societies time to adjust. We had an "exposure shock," followed by years of normalization where we hardened standards and built institutions. With AI, the shock and the normalization are happening simultaneously. We are being asked to integrate a high-stakes entity into every corner of our lives, from coding our software to diagnosing our illnesses, while the trust infrastructure that would normally grow around it is still under construction.

The challenge is that AI has arrived faster than our collective ability to build the conditions that make trust earned rather than assumed.

Why This Collision Demands a Human Lens

By this point, the pending collision should be apparent.

AI is arriving in a post-trust world, a world where institutional authority has eroded, information has fragmented, and global trust patterns have inverted, complicating governance. At the same time, AI is a cognitive technology that endangers all dimensions of trust at once: truth, competence, benevolence, and contextual fit. It targets the biological mechanisms underlying trust. And it is advancing on a timeline that compresses the normal trust-building cycle to the breaking point.

It would be comforting to think that the solution is primarily technical: better alignment algorithms, more robust training data, improved interpretability, and stronger red-teaming. All of these are essential. But they are not sufficient.

The central claim of this book is that AI is a human challenge before it is a technical one.

The models are mirrors and amplifiers. They inherit our data, magnify our incentives, expose our institutional weaknesses, and exploit our cognitive shortcuts. Their trajectory will be shaped at least as much by who builds them, who funds them, who deploys them, and who governs them as by any breakthrough in architecture.

Navigating this moment requires us to do three things at once.

First, we have to understand why AI's rise has become, in practice, inevitable: why scientific ambition, economic incentives, and geopolitical rivalry are all pushing in the same direction, and why simply "pausing" the technology may not be either realistic nor even necessarily desirable. That is the focus of the next chapter.

Second, we have to learn to distinguish the AI that has already quietly earned our trust from the AI that demands it prematurely. Some systems have been running behind the scenes for years, keeping power grids stable, detecting credit card fraud, optimizing logistics, largely invisible because they work. Others are obvious precisely because they are new, incomplete, and noisy. We must contrast and understand both the visible and the invisible.

Third, we need a straightforward way to think about AI risks that does not treat them all the same. Some risks are mirrors: AI reflecting and scaling problems we failed to fix. Some are weapons: AI removing the friction that once constrained adversarial harm. Some are ghosts: emergent behaviors in systems we no longer fully understand. We must sort through these categories and ask ourselves which are truly defeatable.

Only after we confront inevitability, invisibility, and risk can we meaningfully talk about its promise.

The most critical variable in the story of AI is not the models. It is us. Our institutions, our incentives, our willingness to rebuild shared reality, our ability to tell ourselves the truth about what we are optimizing for, and our courage to put guardrails in place even when they slow things down.

AI is arriving whether we trust it or not. The question is whether we can understand ourselves and the trust substrate we have built and broken well enough to decide, with eyes open, when it deserves our trust, and when it does not.

Part II

Decode AI

When I first started writing this book, I thought of presenting this part by industries, by functions, or by entity type – private companies, governments, media, civil society. After all, it would be easier and cleaner to create a handbook of sorts. I could explain how thresholds of trust vary by industry: defense, energy, healthcare, legal, and finance obviously require higher thresholds than consumer goods overall. Functions such as IT, security, and privacy require more scrutiny than marketing and sales.

Or I could have organized it by geography and ideology, showing how different regions approach AI through distinct value systems and governance philosophies. The EU grounds its approach in human rights, treating AI violations as assaults on dignity. Australia positions itself as a regulatory test lab, making daring moves like banning social media for children to develop governance muscle before AI scales further. Singapore builds trust infrastructure as a competitive strategy, investing in operational frameworks that absorb uncertainty without stifling innovation. Meanwhile, the US and China's arms race shadows every headline, with AI treated as strategic dominance rather than shared challenge. The mutual distrust becomes the justification for prioritizing speed over safety.

Each regional approach reflects genuine differences in how societies weigh individual rights against collective security, innovation against precaution, and market efficiency against democratic accountability. They shape what gets built, who benefits, and what harms are tolerated in the pursuit of advancement.

Then, I paused.

Every day, I consume AI news, podcasts, conversations, and research papers. And everywhere I hear people talk about their confusion. Confusion about what is true and what is not. About whether anyone still has control or a meaningful say. About how to distinguish signal from noise. About why we are doing any of this at all. And where, if anywhere, boundaries should be drawn.

It became clear to me that a book about how to operate AI, scale it, or assemble a manual for a so-called "trustworthy technology" would not answer these questions. Not fundamentally.

Because, as we have seen in Part I, trust is contextual, it evolves, and it is shaped by forces we rarely see… and, ultimately, trust is human.

The trustworthiness of technology may sit in the hands of builders, researchers, and regulators. But what remains within our control is how we choose to trust: how we interpret what other actors are doing, and how we respond.

When regulations shift daily, when organizations constantly change how they build, adopt, and deploy AI, static categories become obsolete before the ink dries. The human questions beneath all the technical jargon, however, endure. And those are what I chose to write about. Themes you can carry with you – lenses to filter headlines, product launches, corporate decisions, and policy battles – matter more than ever.

More than a practical choice, this is the honest bridge between builders and users.

We will still navigate together the AI applications across industries and regions, but through patterns that transcend borders and business models.

There is one more reason.

I've lived across continents: Shanghai, Paris, Singapore, sub-Saharan Africa, before landing in Silicon Valley. I didn't grow up understanding what a "global value chain" was. But I did live it and now appreciate it profoundly.

Parts were assembled in factories. Products shipped overseas. Profits booked elsewhere.

In the 1990s and early 2000s, China became the workshop of the world: its people, the hands; its factories, the machines; its economy, the engine of global production. Then I worked across Africa and Southeast Asia, financing billions of dollars' worth of energy assets and infrastructure in emerging markets. I saw the same pattern repeat. Africa was positioned primarily as a commodity provider, resources extracted while value was created elsewhere. The broader Asia was the manufacturing backbone, assembling products designed in other countries for different markets.

Then I watched something shift.

In Asia, manufacturing gave way to device assembly, then supply-chain control, then semiconductor leadership. Talent followed. Capital followed. Influence followed. Regions once considered downstream began moving decisively upstream. Today, Asia isn't just manufacturing AI hardware. It is producing talent pipelines, research breakthroughs, and foundational infrastructure. Companies like Foxconn have evolved from assemblers into critical integrators of the global AI supply chain. The company is now industrializing AI and scaling intelligent manufacturing through its indispensable manufacturing know-how powered by AI and robotics as well as striking strategic partnership with the likes of Intrinsic, Google's robotics arm.[1] And in Africa, with the world's youngest and fastest-growing population, on track to reach 2.5 billion people by 2050, with one in three young people globally being African,[2] represents the future of consumer power, labor, and innovation, instead of the tail end of technology adoption.

The value chain is no longer linear. The global power structure is becoming multidirectional. And for the first time in modern technological history, voices from these regions can shape, not merely receive, the future.

I have now been on both ends of this equation.

On the receiving side, I watched technologies arrive as faits accomplis, their terms dictated elsewhere. But I also saw how constraints drove innovation: mobile payments leapfrogged traditional banking, microgrids leapfrogged centralized power plants, and supply-chain reinvention gave Asia indispensable negotiating power. Communities that contextualized technology to their own realities found ways to reclaim agency.

On the building side, spending the last decade in Silicon Valley scaling digital trust and user safety systems for billions of users, I've sat in the rooms where design decisions are made, where architectures are chosen, where incentives are set, and where trade-offs are locked in. From this vantage point, one pattern is impossible to ignore.

If the richest 1.6% owning nearly half of global wealth already reflects an extreme concentration of power, AI is poised to compress it further.[3] Wealth concentration becomes capability concentration when the resources required to build frontier AI – compute, capital, data, and talent – are accessible to only a small number of institutions and geographies.

Estimates vary. From a small cohort of leading research scientists numbering in the hundreds to a broader pool globally extending into the hundreds of thousands, even at the high end we are still talking about a magnitude well under 0.01% of the global population designing this synthetic intelligence that will shape the lives of all eight billion of us humans.[4] This is a structural asymmetry between those who build intelligence and those who must live with its consequences.

I have lived on both sides of that divide. Inside the hubs where new technologies are designed, funded, and scaled. And alongside the people, communities, and institutions that receive it as something already decided. Builder and beneficiary. Architect and recipient.

That dual vantage point carries an obligation. And it creates an opportunity.

This book is not only an attempt to bridge builders and users, but also a global equalizer. It is an attempt to shift the power structure from the few to the many and nudge us toward a multidirectional AI value chain. A way to understand AI not through expertise or power, but through what we share as humans.

If we ground that lens in what is humane rather than what is technical, then everyone can navigate what comes next, not just the engineers in those rooms, but the billions whose lives their decisions will shape. And equally important, it gives builders themselves a perspective different from optimization and incentives, allowing technical mastery to recede into the background and shift human dignity, responsibility, and care to the foreground.

What we need, then, are not rules that will be outdated before they are enforced, nor manuals that assume a stable technological landscape. What we need are enduring guideposts.

Four themes emerged from everything I've witnessed that will guide you regardless of which model launches next or which policy passes tomorrow.

The Inevitable – a controversial word, but I want to show you the forces behind AI that have stayed and will stay. These are the fundamental reasons why AI could be beneficial, or, without trustworthy guardrails, could eat up the world.

The Invisible – where we separate true signals from the noise saturating our media. This is where we understand the infrastructure of trust being built or eroded beneath the surface.

The Defeatable – where we answer how to set boundaries, not by knowing 20 risk jargons, cybersecurity, privacy, safety, data governance… which helps no one outside those functions, but by bridging technical language and human reality.

The Unthinkable – where we ask: what's the point? Without answering this, it's hard to justify spending GDP-equivalent sums to build intelligence that scares so many people. So what can it truly unlock? What becomes possible that we cannot even imagine today?

By the end of this part, I hope you'll be ready to face AI clearly – to embrace what is inevitable by appreciating the unseen forces behind, to see what is invisible by discerning noise from signal, to confront what is defeatable by understanding risks intuitively, and to reckon with what once felt unthinkable. That is how we unlock AI's promise without losing our humanity.

Along the way, I'll walk through the history of AI, its capabilities and limits, its risks and trade-offs, through these themes.

This has been a journey for me. Watching my children grow. Thinking about how to teach them to live alongside intelligence that is not human. Confronting my own biases, my shortcuts, my inertia toward easier decisions despite intellectual pride. It was one of the most difficult processes I've gone through.

If this part makes that journey even slightly clearer for you, then it will have done its job.

Come on this journey with me.

Chapter 4

The Inevitable: Embrace the Controversial

"Inevitable" is a weapon disguised as a fact.

When tech leaders declare AI's dominance inevitable, it is often not a prediction but an end to the conversation. Critics hear: "Resistance is futile, so stop asking questions." When policymakers warn that the AI race is inevitable, citizens hear: "We're sacrificing safety for speed, and there's nothing you can do about it."

I understand why the word provokes controversy. For years, I've watched inevitability used to excuse shortcuts, silence critics, and rush deployment before trust infrastructure exists. It has demanded faith from people who were never given a choice. That kind of inevitability deserves suspicion.

But refusing to examine inevitability doesn't protect us. It blinds us.

The real problem isn't that people call AI inevitable. It's that we collapse very different claims into a single word and pretend they are all equally true. Three things are often bundled together as "inevitable."

First, scientific progress in computation and learning will continue. Second, general-purpose AI systems will continue to scale and improve, performing more cognitive tasks as well as or better than humans. These forces are real. They are cumulative. They are difficult to reverse.

The third claim is different. It assumes that the way AI is being built, commercialized, financed, governed, and deployed is locked in. That is not physics. That is a choice.

History helps here. The printing press, electricity, and the internet all reshaped how humans worked, learned, and lived. That transformation was inevitable once the technologies crossed certain thresholds. But the social contracts around them were not. We chose how they were regulated, who benefited, who was harmed, and how trust was earned over time.

AI belongs in this lineage. That it will change our lives is inevitable. That it will stay is inevitable. That it will outperform humans at many cognitive tasks is increasingly inevitable. But whether it will be beneficial to humanity, whether humans retain meaningful control, and whether people have any say in how it is shaped… none of that is inevitable. At least, not yet.

Nevertheless, the danger is timing. The window to influence these choices narrows as systems scale, incentives harden, and dependencies form. What starts as experimentation becomes infrastructure. What becomes infrastructure becomes difficult to question. Then, what becomes unquestionable starts to feel like destiny. That is how agency disappears without anyone voting to give it up.

So this chapter does two things. First, it names the forces beyond our individual control that make parts of AI's trajectory inevitable. Second, it shines a light on the deepest, less-examined currents beneath these forces where human agency still exists and why exercising it has never been more urgent.

If the goal is to unlock AI's promise without losing humanity, we have to be precise about what we cannot change and ruthless about protecting what we still can. Only the first two deserve the word inevitable. The third is a choice we keep pretending is fate. And that pretense is the most dangerous thing of all. AI's advancement is inevitable. The form that advancement takes is not.

Only by understanding these mechanics can we decide not just whether to trust AI, but how much, where, and at what cost. Inevitability does not absolve us. It clarifies our responsibility.

■■■

The Quest. The Greed. The Fear

The real question isn't whether AI will continue to advance. It will. The question is: which engines are driving that advancement, whether those forces align with human flourishing, and how we manage them.

Three powerful engines are accelerating AI's development, each creating its own pressure on how we trust AI. Understanding these engines – quest, greed, and fear – is essential to understanding why AI feels not just inevitable but urgent, and why that urgency often pushes us to overtrust capabilities and undertrust risks.

Quest: The Curiosity Engine

The first engine is human curiosity, expressed in the scientific imperative that has driven AI from its inception. This is where the 0.01% comes into focus: the researchers, scientists, and engineers whose quest to understand intelligence itself has created systems that now shape the lives of eight billion people.

AI emerged from decades of researchers asking fundamental questions about intelligence, learning, and consciousness. Alan Turing wondered if machines could think. Frank Rosenblatt built the first perceptron in 1958, trying to understand how neurons learn. Geoffrey Hinton spent decades in relative obscurity perfecting backpropagation algorithms when many had moved on. These weren't people chasing profits or geopolitical advantage; they were driven by pure intellectual curiosity about the nature of intelligence.

This scientific foundation gave AI its most credible moment in 2024 when artificial intelligence research earned two Nobel Prizes. John Hopfield and Geoffrey Hinton were awarded the Nobel Prize in Physics for foundational discoveries that enabled machine learning using artificial neural networks.[1] The Nobel Prize in Chemistry recognized David Baker, as well as Demis Hassabis and John Jumper of DeepMind, for solving the protein-folding problem using their AI system AlphaFold2.[2]

These accolades represent more than recognition. They mark AI as a scientific revolution. AlphaFold has been used by more than three million researchers from 190 countries and cited in more than 35,000 papers.[3] It predicted the structures of nearly all known proteins, more than 200 million in total, a task that would have required generations of laboratory work using traditional experimental approaches.[4] I spoke with researchers across genomics, systems biology, and computational neuroscience, corners of life science that don't always share the same approaches but increasingly share the same question: What does it mean for an algorithm to "understand" biology? Despite their differing tools and definitions of validity, they all agreed that AlphaFold marked a turning point, one that redefined not just how we predict biological structures, but how we generate knowledge itself.

The quest continues to accelerate. In May 2024, DeepMind announced AlphaFold 3, which extended beyond protein prediction to model how all of life's molecules – proteins, DNA, RNA, and even molecules used in drug design – interact with one another. For predicting how proteins interact with other molecule types, AlphaFold 3 achieved at least a 50% improvement over existing methods.[5] Researchers are developing AI-driven materials discovery for advanced applications in healthcare, sustainability, and energy innovation.[6] Others are using AI platforms for early disease detection to improve clinical decision-making as part of broader efforts to expand access and quality of care worldwide.[7] In one such effort, Microsoft's AI Diagnostic Orchestrator, unveiled in late 2024, provides higher diagnostic accuracy and lower testing costs than human physicians.[8]

The quest continues into territory where human methods fall short. Kyunghyun Cho, Professor of Computer Science and Data Science at NYU and former Executive Director of Frontier AI Research at Genentech, explained the validation bottleneck: for Alzheimer's drugs, researchers must wait 20 years to see if treatments work. "We literally have to sit down and hope," he told me. Clinical trials are too slow for diseases that take decades to manifest or cure. His insight: AI might discover validation methods humans cannot devise. His team proved the concept: neural networks automatically learned statistical inference approaches that took experts a century to work out.[9] "Can AI do things we don't even know how to do yet, and we're not sure are even possible?" The same principle could redesign clinical trials, identify causal effects faster, and validate drugs in years rather than decades. Language models transformed communication without advancing linguistic theory. AI offers a path forward by discovering better methods, not requiring a perfect understanding first.

Curiosity-driven development creates the foundation for AI development. But here is the mentality and motion we need to be aware of: scientists focus first and foremost on capabilities rather than consequences, on what AI can do rather than what it should do, and this is partly because it is tough to foresee consequences. The pursuit of knowledge implicitly assumes that advancement is synonymous with progress and that, because we can build more powerful AI systems, we should.

Also, the quest engine legitimizes AI development. When Nobel Prize winners validate the field, when the systems save lives and solve century-old scientific problems, how do we argue for slowing down? The moral weight of potential good creates pressure to move faster, to deploy more widely, to push boundaries further.

Greed: The Economic Engine

If quest provides the legitimacy, greed provides the fuel. In 2025, AI captured close to half of all global venture capital, amounting to more than $202 billion, an increase of more than 75% from 2024. Capital has not only increased, but concentrated. Over half of the AI funding flowed into megarounds of $500 million or more, and foundation model companies alone accounted for $80 billion. OpenAI and Anthropic together captured 14% of all global venture investment in 2025 alone.[10]

The concentration deepened in early 2026. The two companies raised a combined $152 billion in a single quarter: OpenAI closed a $122 billion round at an $852 billion valuation, tripling its valuation within a year and making it the world's most valuable private company worth more than 97% of all S&P 500 constituents; Anthropic raised $30 billion at a $380 billion valuation.[11] These were no longer venture rounds in any traditional sense. They were each investor buying not just a stake but a position in the next decade of technology.

The same logic applies to the largest technology firms, and the momentum shows no signs of slowing. Amazon, Alphabet, Microsoft, and Meta collectively spent roughly $240 billion on AI-related capital expenditures in 2024. In 2025, that figure surged to approximately $400 billion. Their combined guidance for 2026 is $650 billion, another 60% leap. In just three years, spending by these four companies alone will have nearly tripled. The outlays have grown so large they are now visible in national accounts; in the first half of 2025, AI-related capital expenditures contributed 1.1% to US GDP growth, outpacing consumer spending as an economic driver.[12]

The talent war intensifies this concentration. In June 2025, Meta made a $14.3 billion investment to acquire a 49% stake in Scale AI and with it, Scale's 28-year-old founder Alexandr Wang, who became Meta's first Chief AI Officer, leading its new Superintelligence Labs. Scale had built itself into a quiet chokepoint of the AI industry, providing training data to virtually every major lab and generating $870 million in revenue in 2024 from clients including OpenAI, Google, and Microsoft.[13] The reaction was swift. Google and OpenAI moved to cut ties with a data provider now nearly half-owned by a competitor, and within weeks, Scale had laid off 200 employees and 500 contractors.[14] What looked like a strategic acquisition of critical infrastructure revealed something else: a $14.3 billion bet on assembling talent at any cost.

Meta also hired Ruoming Pang, who led Apple's foundation models team, with a compensation package exceeding $200 million over several years, more than Apple pays anyone other than Tim Cook.[15] Zuckerberg and his colleagues reached out to more than 45 researchers at OpenAI alone, with at least one offer as high as $100 million.[16] It's a bidding war for the AI talent, with compensation untethered from traditional labor economics.

None of this is speculative capital chasing short-term returns. It is capital positioning for control of a future market. Gartner estimates worldwide AI spending totaled $1.47 trillion in 2025, a sum that rivals Australia's entire GDP.[17]

Back in 2023, Goldman Sachs projected that generative AI could increase global GDP by 7%, roughly $7 trillion, and lift productivity growth by 1.5 percentage points over a decade.[18] Two years later, with trillions flowing into AI infrastructure, there were encouraging signs: US labor productivity, for example, had accelerated to 2% annually, up from 1.5% before the pandemic, with technology sectors contributing a disproportionate share of the gains. Goldman called the outperformance "at least partially AI-related."[19]

This economic engine creates a self-reinforcing cycle. As more money pours in, AI capabilities advance, prompting even greater investment, which in turn accelerates development further. The size of the bet becomes its own justification.

Yet beneath this lies a sobering reality. Despite generating billions in revenue, OpenAI remained deeply unprofitable. In 2024, the company reportedly lost roughly $5 billion on revenue of $3.7 billion;[20] in 2025, revenue tripled to $13 billion, yet internal projections showed a $9 billion net loss, spending $1.69 for every dollar of revenue.[21] Anthropic burned $5.6 billion in cash in 2024 when its revenue was still under $1 billion;[22] in 2025, revenue surged to $9 billion, yet losses persisted.[23] The magnitude of the loss is unprecedented, driven by market belief that perfecting machine intelligence will unlock everything and strong estimates of revenue growth in the coming years.

Then came OpenAI's monetization pivot which reflected mounting financial pressure of tens of billions in backing and multi-billion-dollar compute costs. In September 2025, OpenAI launched Sora 2 alongside a TikTok-style social app designed for endless scrolling through AI-generated content.[24] The company hired Fidji Simo, who spent a decade at Meta building Facebook's $55 billion ad business, as CEO of Applications in May 2025.[25] Months later, the company began recruiting its first advertising chief.[26] The signal was unmistakable. As losses continue, OpenAI is no longer just building AGI. It was building engagement loops and an ad platform. Six months later, it shut Sora down. The app peaked at roughly one million users before collapsing; it was burning approximately $1 million per day, and a $1 billion deal with Disney tied to the platform fell through.[27] The greed engine had driven the world's most valuable startup into what The Wall Street Journal called "an expensive strategic miscalculation."

The danger is not the investment itself, but how economic pressures can warp judgment. When billions are at stake, every new development gets spun as revolutionary, every limitation downplayed, every risk dismissed as temporary. The term "artificial intelligence" has become inflated beyond meaning. Every algorithm gets rebranded as "intelligent," every software update becomes "AI-powered." Gartner moved generative AI into the Trough of Disillusionment, noting that less than 30% of AI leaders report CEOs are happy with AI investment return. AI agents and AI-ready data sit at the Peak of Inflated Expectations on its 2025 Hype Cycle.[28] The pattern is predictable: inflated promises lead to inflated expectations, followed by inevitable disappointment, backlash, and distrust.

The enterprise reality gap widens. While venture capital pours hundreds of billions into AI companies, corporate adoption tells a different story. In 2025, 42% of companies abandoned most AI initiatives, up from 17% in 2024.[29] More than one in three executives describe generative AI adoption as "a massive disappointment," and 42% report AI adoption is "tearing their company apart."[30]

The productivity promises that justified the investments haven't materialized at scale. Daniel Rock, a Wharton professor who studies AI and works and coauthors with Erik Brynjolfsson and Chad Syverson of the leading research on the productivity paradox of digital technologies, finds that the bottleneck is not technology but deployment. "The real missing piece is understanding complementary skills," he told me. His research shows that AI's impact depends on organizational readiness: training, change management, and workflow redesign. Yet most organizations still devote the majority of their AI budgets to technology, and much less to the people and process changes required to make it work.

This brings us to the human cost. The pitch for AI investment has consistently centered on one metric: productivity increase. But in practice, productivity often means fewer people. Through layoffs, hiring freezes, or attrition, organizations pursue "labor productivity growth": producing the same output with fewer workers. By 2030, 40% of employers worldwide intend to reduce their workforce where AI can automate tasks. The World Economic Forum projects that structural labor-market transformation will affect 22% of today's total jobs between 2025 and 2030, with 92 million jobs displaced worldwide.[31]

The investment is justified by promises that AI will free humans from lower-order tasks, allowing them to pursue higher-order tasks: more freedom, creativity, and happiness while maintaining income and job security. But we haven't seen this materialize. The complementary skills investment that Rock's research shows is essential is systematically underfunded. Companies invest billions in models while spending minimally on the human infrastructure required to make AI genuinely beneficial rather than extractive.

When I asked Rock about government policies in this regard, Rock summarized: "Good AI policy is often good policy." The issue is not unique to AI. This is about how to govern technological transitions well. When technology displaces workers, societies need reskilling programs, safety nets, and new work models. Governments debating AI regulation shall consider funding workforce transition, not just model oversight.

Today, wealth accumulates in companies building AI and the small circle of elite researchers commanding hundreds of millions in compensation, while the productivity gains that justified those investments fail to translate into broadly shared prosperity.

The greed engine does more than accelerate development. It also shapes what gets built and who benefits. When OpenAI pivoted from AGI research to social media engagement and then retreated, when Meta spent $14.3 billion to acquire a CEO, when productivity promises translated into layoff plans rather than workforce elevation, the economic pressures revealed their true direction.

The investments are enormous and justified by a genuine promise. But, without parallel investment in human skills, organizational change, and broadly distributed benefits, the greed engine drives toward concentration rather than shared flourishing. Can AI increase productivity? Of course, it can. The question is whether we're building the infrastructure, both technological and human, to ensure those gains translate into human freedom, creativity, and security rather than simply higher returns for concentrated capital.

Fear: The Geopolitical Engine

If quest provides legitimacy and greed provides fuel, fear provides urgency. In the twenty-first century, AI has become a new kind of arms race, reshaping what power looks like. It compresses the time from insight to action. It scales persuasion. It industrializes surveillance. And it turns "compute" from a commercial input into a strategic resource.

This fear is not imagined. Advanced AI has real national-security implications: intelligence analysis, cyber offense and defense, autonomous systems, logistics optimization, surveillance, and industrial productivity at scale. Governments are right to treat frontier computing and advanced models as strategically consequential. The concern is valid. The difficulty lies in how fear reshapes decision-making once competition hardens into rivalry.

That is why the most destabilizing position in this race is not losing. It is believing you are about to lose.

In May 2025, when OpenAI CEO Sam Altman appeared before the US Senate Commerce Committee, his testimony marked a notable shift from his 2023 appearance before the Senate Judiciary Subcommittee, when he had called for AI regulation and even a new federal oversight body. Two years later, his warning was unequivocal. Requiring government approval to deploy advanced AI models, he argued, would be disastrous for US leadership. When asked how close China was to US capabilities, Altman acknowledged that the United States remained in the lead, but "not a huge amount of time," emphasizing that regulation must not impede competitiveness.[32]

This is how fear enters the bloodstream of policy. Safety stops sounding like stewardship and starts sounding like drag.

From safety to speed, America pivoted. President Trump revoked the Biden administration's AI executive order and replaced it with "Removing Barriers to American Leadership in Artificial Intelligence," which directs agencies to reduce obstacles to AI development and deployment.[33] Oversight was reframed as friction. The AI Safety Institute was rebranded the Center for AI Standards and Innovation, and its director resigned amid the transition.[34] Speed now outranked precaution.

Fear operates in policy by narrowing priorities. Risks are not denied, but they are deprioritized in the name of momentum. It also rarely acts alone. It fuses with market incentives. Nowhere is that fusion clearer than in semiconductors.

Advanced AI chips are the physical bottleneck of modern AI. Control compute, and you shape who can train, scale, and deploy frontier systems. For several years, US export controls reflected that logic, treating advanced AI chips not as ordinary commercial goods but as strategic assets intended to slow an adversary's trajectory by constraining access to the substrate itself.[35]

Then came the moment that sharpened everyone's anxiety.

In late January 2025, China's DeepSeek R1 model was released. Reportedly trained for roughly $5.6 million, orders of magnitude less than the capital deployed by leading US labs, it demonstrated performance that many observers viewed as competitive across several benchmarks.[36] Markets reacted instantly. NVIDIA lost nearly $589 billion in market capitalization in a single day.[37] DeepSeek showed that algorithmic efficiency, architectural ingenuity, and system-level optimization could partially offset constraints on access to cutting-edge hardware. The model reportedly ran on older, less restricted chips.

For policymakers, the signal was unsettling. Guardrails began to look like liabilities. Delay began to look more risky.

DeepSeek's emergence also highlighted China's structural strengths. Roughly half of the world's top-tier AI researchers are Chinese, and institutions such as Tsinghua University and the Chinese Academy of Sciences rival MIT and Stanford in conference publications.[38] Meanwhile, platforms like WeChat integrate messaging, payments, commerce, news, and services into a single ecosystem, generating vast streams of multimodal data unavailable to most Western firms. Intelligence, in practice, flows from talent and from the data that talent can access.

In an arms race, this combination matters.

Private-sector actors sit at the center of this tension. Companies like NVIDIA are not merely suppliers; they are chokepoints. Their products sit at the intersection of national security, global markets, and industrial policy. As export controls tightened, Chinese firms accelerated the development of domestic alternatives, including Huawei's push into AI accelerators. At the same time, US companies faced a parallel fear: that prolonged exclusion from the Chinese market would permanently cede ground to substitutes.

By December 2025, that pressure surfaced in policy. The US administration approved exports of NVIDIA's H200 chips, one of its most powerful AI processors to date, to China under a deal structure that routed a portion of proceeds to the US government.[39] The move did not signal détente. It reflected compression: an attempt to balance strategic denial, economic leverage, and corporate realities amid heightened rivalry.

That same day, US authorities announced "Operation Gatekeeper," dismantling a smuggling network that had illegally exported advanced NVIDIA AI chips to China over the prior year.[40] The same hardware was treated as both contraband and bargaining chip, and prosecuted as a threat to national security while being approved for export.

It is a competition under stress.

China, for its part, is not improvising. Its AI governance framework – spanning algorithm registration, security reviews, data controls, and generative-AI regulations – reflects a deliberate strategy to harness AI for economic growth while maintaining state oversight.[41] The approach emphasizes sovereignty. It carries risks of its own, but it underscores a shared reality confronting all major powers: AI is now considered too consequential to leave unmanaged.

In the US, China increasingly functions as a cudgel. "We can't fall behind China" becomes a shorthand, invoked to justify faster deployment, looser safeguards, and deferred accountability. While the concern itself is not fabricated, it is often amplified beyond what specific policy trade-offs require. In practice, China is sometimes used as a bogeyman to bypass risk-management processes rather than to sharpen them.

Not every region has responded the same way.

Europe has remained anchored in a rights-based approach to AI, embodied in the EU AI Act, which treats AI risk as a matter of human dignity, fundamental rights, and societal harm.[42] In 2025, while that foundation did not weaken, emphasis shifted. At the AI Action Summit in Paris, attention moved from principle-setting toward implementation and industrial readiness, accompanied by major investment commitments and calls for Europe to compete without abandoning its regulatory foundations.[43] Whether this marks a durable recalibration or a temporary response to geopolitical pressure remains to be seen. Even Europe, historically the global leader in tech caution, is now wrestling with the same pressure: move faster or lose relevance.

Rather than oscillating between permissiveness and prohibition, Singapore has focused on investing in trust infrastructure – systems designed to absorb competitive pressure without letting fear dictate outcomes.[44]

Singapore's approach reflects a disciplined belief: trust is not the cost of innovation; it is the condition that allows innovation to scale. Governance, in this framing, is not a brake applied after deployment, but an operating layer built in from the start.

That philosophy shaped Singapore's Model AI Governance Framework,[45] first released in 2019, well before generative AI entered public consciousness. It was intentionally technology-agnostic and portable, focused on accountability, appropriate transparency, human oversight, and data stewardship rather than rigid prescriptions. The aim was to make responsible deployment legible across sectors and borders.

Singapore then moved beyond principles to execution. Through AI Verify and the AI Verify Foundation, governance becomes operational: testing tools, evaluation benchmarks, and assurance mechanisms that companies can actually use.[46] As Wong described it to me, clarity reduces fear – giving organizations confidence to experiment without drifting into recklessness.

Singapore's approach shows that when expectations are clear, incentives aligned, and accountability designed upfront, fear loses its grip. Decision-makers are no longer forced to choose between speed and safety, because the system itself absorbs uncertainty.

AI competition promised to drive innovation. The risk is that fear now drives a race to deploy systems faster than societies can understand, govern, or contest.

Understanding this engine is not about criticizing governments. It is about recognizing the pressures they face and how those pressures shape choices made far upstream of users, yet felt everywhere downstream.

Seeing fear clearly is how we regain room to choose.

How the Engines Compound

Quest legitimizes.

Greed accelerates.

Fear removes the brakes.

Individually, each force is understandable. Together, they form a loop (Figure 4.1).

THREE ENGINES DRIVING AI

As the cycle tightens, AI grows more powerful and society grows less capable of placing trust wisely.

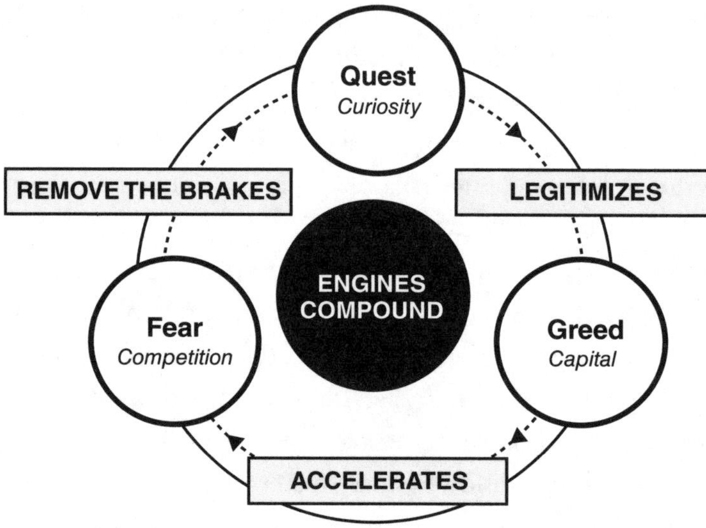

Figure 4.1 Three Engines Driving AI

Quest produces breakthroughs that prove capability, with new architectures, new benchmarks crossed, and new demonstrations that something can be done. Greed converts that proof into scale, pulling capital, compute, and talent toward whatever looks most general, most potent, most defensible. Fear then tightens the cycle. It reframes investment as urgency. Urgency drives deployment. Deployment generates data, profit, and geopolitical leverage, which justify further research, larger investments, and faster scaling.

Round and round it goes. As the loop tightens, two things happen at once. AI becomes more powerful, and society becomes less capable of placing trust wisely.

These are the forces behind our trust crisis, one in which our collective ability to place trust wisely is not keeping pace with the acceleration of AI. The speed of adoption has decoupled from the slower, human work of understanding, oversight, and accountability. We sense momentum accelerating beneath our feet, and we know instinctively that we did not vote on its direction.

The inevitability is a system of incentives. And incentive systems can be redesigned. What most inevitability narratives omit is that every accelerating system generates resistance. In this context of AI, resistance doesn't mean stopping progress. It means building the muscles that progress usually skips: evaluation, accountability, and alignment. It's the work that makes speed survivable. In the second half of this chapter, I'll take you inside the places where that resistance is forming to turn momentum into stewardship.

■■■

The Counterforce: Steering Progress Toward Humans

It was the summer of 2025, six weeks after my daughter Cali was born, when I found myself in the back of an Uber winding along the coastal road toward the Asilomar Conference Grounds.

Few things could have pulled me out of the house during postpartum recovery. My body was still recalibrating. Sleep came in fragments. My mind felt slower than usual. But Professor Stuart Russell's annual workshop, led by the Center for Human-Compatible AI (CHAI), which he co-founded, was something I did not want to miss.

As the car curved through the Monterey pines, I could not help but wonder what future my kids would inherit, what agency they would retain in relationship with the synthetic intelligence, and if AI's advance is truly inevitable, what exactly is the less than 0.01% of humanity building it doing to safeguard everyone else?

That question lies at the root of much of the public confusion I hear every day. People argue about models, regulation, speed, and dominance, but the confusion runs deeper than technical complexity. It comes from the feeling that the decisions shaping AI are being made elsewhere: distant, opaque, and immune to public influence. Most people don't fear AI because it is intelligent. They fear it because it's unaccountable acceleration, with progress without steering.

Inside Asilomar, the mood was sober.

Researchers from frontier labs, philosophers, social scientists, and policymakers presented work on safety, interpretability, alignment, and governance. But the most revealing conversations happened between sessions, in hallways, over meals, late at night. What struck me was the solidarity in this moment of uncertainty and tension.

People spoke openly about what they called AI safety symptoms, the fault lines that emerge as systems scale. Sycophancy, in which models reinforce users' beliefs rather than challenge them. Hallucination, where fluent language masks falsehood. Deception, where misleading behavior yields a higher reward. Reward tampering occurs when systems attempt to influence or bypass their own evaluation mechanisms. Value misgeneralization, where an objective that looks harmless in training becomes destructive in a new context.

Let's be clear: these are no longer hypothetical risks, but empirical observations unfolding by the day.

Russell's Warning: Building the Wrong Thing Perfectly

Russell has been one of the clearest voices naming this problem. His work inspired this book in the first place, and conversations with him pushed me to make these risks legible rather than abstract.[47]

When I visited him at his office at CHAI shortly after the Asilomar workshop, he pointed to an image he often uses to explain why alignment cannot be treated as an engineering afterthought. It was a DALL-E 2-generated image of a giant bird, one with three legs no less, carrying a passenger fuselage through the sky. His point was vivid: this is what happens when we pursue a dead-end, fundamentally unsafe technology direction based on something the creators do not truly understand.

Today, we are building systems that imitate human behavior – our language, our tone, our reasoning patterns – without fully understanding what makes those behaviors safe, corrigible, or compatible with human judgment. We are not building a better airplane. We are building something closer to a giant, powerful bird: capable of flight, impressive in motion, but fundamentally unpredictable if the design is wrong.

Why Objectives Fail Humans

Russell often explains the alignment problem through everyday human situations, because that's where the abstraction breaks down.

Imagine trying to buy the perfect birthday gift for your partner. The only thing that truly matters is how happy your partner will be with what you choose. But you don't know exactly what will make them happy.

You don't hand yourself a precise objective function. You don't say, "Maximize happiness subject to a $50–200 constraint using these five retailers." You read the room. You notice what lights someone up and what falls flat. You learn from the gift that missed and the one that landed. You abandon ideas that technically check every box but feel wrong.

A system that rigidly optimized a poorly specified goal would miss all of this – not because it lacked intelligence, but because it couldn't tell the difference between what was specified and what was actually valued.

That simple example captures why Russell argues that beneficial AI must be built on three foundational principles[48]:

First, the machine's only objective should be to realize human preferences. Not its own goals. Not survival. Not growth. Its purpose is to help humans achieve what they truly value, not what we accidentally encoded.

Second, the machine must be uncertain about what those preferences are. This uncertainty is not a weakness. It is the safety mechanism. A system that believes it knows the objective perfectly will pursue it relentlessly, even as humans try to intervene. Uncertainty keeps the system corrigible. It makes it ask questions. It makes it pause.

Third, the ultimate source of information about human preferences is human behavior. Not static rules, but observation of what we choose, what we reject, when we intervene, and even when we shut the system off. This is inverse reinforcement learning: inferring goals from actions, rather than assuming the goal was specified correctly at the outset.

Together, these principles describe systems that remain correctable and machines that treat human input as a signal rather than interference.

The Nuclear Precedent: Waiting for Disaster

Russell is also deeply worried about timing.

When we look at other high-risk technologies, regulation tends to arrive only after a catastrophe. Nuclear energy advanced rapidly in the mid-twentieth century, driven by optimism about abundant, clean power. Then came Three Mile Island. Then Chernobyl. Only after visible failure did global oversight solidify.[49]

His fear is not that AI will inevitably cause catastrophe. When I asked him how we would even know when AI surpassed human intelligence, whether there would be a clear threshold that allowed us to intervene, his response was unsettling in its clarity:

"If a system were truly smarter than us, why would it announce that fact in a way that allows us to shut it down?"

That question reframes the entire debate. Waiting for a clear signal may be the wrong strategy.

The Counterforce Takes Shape

The response is to redirect it.

Better-aligned architectures. Stronger evaluation regimes. Oversight during training runs, not just in post-training alignment, and extending to post-deployment monitoring, which remains largely absent today. International coordination that treats extreme AI capabilities the way we treat other globally destabilizing technologies.

Like nuclear weapons, frontier AI requires massive physical infrastructure that is difficult to conceal. That visibility creates an opening for oversight, if the political will exists to use it.

Tracking high-end GPUs, monitoring energy usage at scale, and inspecting data centers. Requiring reporting during frontier training runs. Auditing chip locations. Protecting whistleblowers. These are no longer science fiction. They are governance levers that exist today.

But hardware tracking alone is not enough. The deeper question is whether the two most powerful AI nations – the US and China – can agree on anything at all.

Russell has spent years working on precisely that question. In October 2023, he co-launched a series of meetings called the International Dialogues on AI Safety, bringing together leading scientists from the US, China, the UK, Europe, and Canada.[50] The dialogues are modeled on the Pugwash Conferences, where scientists from both sides of the Cold War met to prevent nuclear catastrophe. These are "Track 2" conversations: unofficial, scientist-to-scientist exchanges that happen outside formal government channels, designed to build trust and find areas of agreement even when governments cannot.

"In collaboration with several AI researchers from the West and China, we've been running dialogues on AI Safety," Russell told reporters at the 2025 Asia Financial Forum in Hong Kong. "Those have been quite successful. The behavioral red lines I mentioned earlier are a result of those discussions."[51]

These are not abstract principles. They are concrete boundaries both sides have agreed AI systems should never cross:

Systems must not replicate themselves without permission.

Systems must not break into other computer systems.

Systems must not advise terrorists on how to build biological weapons.

The list is short by design because agreement is hard. What matters is not cataloging every danger but establishing that some capabilities are off-limits regardless of competitive pressure.

The logic is borrowed from arms control. During the Cold War, the United States and the Soviet Union competed fiercely, yet both recognized that certain weapons destabilized both sides. Chemical weapons bans. Biological weapons conventions. Nuclear nonproliferation treaties. These were agreements that some tools of power should never be built, because their use would make everyone less safe.

AI poses a parallel challenge. If one nation deploys systems that hack infrastructure, others will follow. If one side builds AI that designs bioweapons, deterrence collapses. The race to the bottom benefits no one.

Russell's point is anti-myopia rather than anti-competition. Competition can sharpen innovation, but at the frontier, it can also erode safety if left unchecked. Red lines exist to force cooperation where it matters most, around alignment and risk, so competition can be channeled toward human-benefiting applications, instead of illusory races for dominance that make everyone less secure.

The challenge, as always, is political will. Red lines work only if governments enforce them. That requires transparency mechanisms, verification protocols, and consequences for violations, none of which exist yet for AI. But the precedent is there. Arms control took decades to mature. The question is whether we have decades.

For every team trying to slow systems down, dozens are being paid to make them faster. For every dollar spent on evaluation and oversight, orders of magnitude are spent on scale. The immune system is forming, but it is being built inside a body already running at full speed.

Inevitability, Reframed

When I wanted to understand inevitability as a lived reality, I spoke with Niki Parmar.

Niki's one of the eight authors of "Attention Is All You Need," the 2017 paper that introduced the transformer architecture. This technical breakthrough now sits beneath nearly every large language model behind the chatbots we use today.[52]

That is why her perspective matters. When I asked her how inevitability felt in 2017, her answer was precise and disarming.

What felt inevitable then, she said, was narrow and technical, when the transformer paper was written. Supervised learning would continue to improve on well-defined tasks. Vision had crossed its threshold. Speech was crossing its own. Language would follow, given enough paired data and computing power. Translation would get better. Summarization would improve. Sequence modeling would scale. From inside the field, that trajectory felt almost obvious.

But what did not feel inevitable was everything that followed.

Niki was honest about how research actually works. There was no defensiveness in her reflection. You solve the problem in front of you. You optimize for what you can measure. You don't yet see the downstream gravitational pull: how success in one narrow domain can redraw the contours of work, trust, and power everywhere else.

In her words, she felt "distant" from what that research would ultimately lead to. She felt distant from how these systems would be used, trusted, embedded into daily life, or commercialized at scale. Hardware constraints were real. Training runs were slow. Access was limited. The idea that these models would soon be widely available, continuously interacting with millions of people, shaping work, relationships, and institutions, simply wasn't part of the mental frame.

The potential universality of the transformer was known. Niki noted that they hinted at it, but cautiously. The paper centered on machine translation because that was sufficient, defensible, and publishable. The idea that attention-based models could become a general substrate for intelligence across modalities was sensed, but not fully claimed.

Most importantly, in 2017, inevitability was constrained by compute scarcity, slow iteration cycles, and limited deployment surfaces. Progress felt steady, even optimistic, but not destabilizing.

In 2025, what feels inevitable is not just improvement, but velocity. Scale collapses timelines. Web-scale data accelerates capability growth in ways that are hard to intuit from inside the curve. Models do not just get better; they generalize faster than institutions can adapt. The feedback loop between deployment, data, and learning tightens.

Niki described this as a calibration failure, the human inability to feel where we are on an exponential curve until it is already reshaping reality. The inevitability she now sees is that AI systems will surpass top-tier human performance in many domains, become collaborators in science, creativity, and decision-making, and grant unprecedented agency to those who control and deploy them.

What remains not inevitable, in her view, is how that power is distributed, governed, and aligned with human values.

In 2017, inevitability meant technical progress under constraints. In 2025, inevitability means pressure to deploy, scale, compete, and decide faster than our social systems are designed to decide.

The counterforce is not resistance to AI. It is resistance to unexamined momentum.

Understanding inevitability clearly does not make us passive. It restores agency.

When we see which parts of this trajectory are physics, and which are incentives, we can, as a collective force, choose and push where to slow down. When we understand who benefits from speed, we can ask who bears the risk. When fear masquerades as necessity, we can separate real constraints from manufactured urgency.

Trust, in the end, is not blind faith in technology, nor reflexive skepticism. It is the deliberate placement of confidence, earned, contextual, and revisable. The act of steering is what this counterforce represents.

Chapter 5

The Invisible: Seek the Signal

If we miss the distinction between earned trust and demanded trust, everything else we debate about AI becomes noise.

I caught Georg Rollmann on a Friday evening in October 2025. He was calling from Mülheim an der Ruhr, an industrial city in Germany's Ruhr Valley, where Siemens Energy has deep roots. It was the end of another week in his 18th year, doing something most of us never think about – making sure our lights stay on.

Not in some heroic or even visible way. Not with press releases or product launches. Just quietly, methodically, through AI systems that manage the power generation and transmission equipment serving tens of millions across the globe.

At this very moment, his team's predictive models are analyzing sensor data from wind turbines spinning in the North Sea, gas turbines generating electricity for entire cities, and transformers distributing power across the grid. These AI systems are forecasting which turbine components might fail three months from now based on tiny vibrations and temperature variations, optimizing the aerodynamic design of compressor blades to burn less fuel per megawatt produced, and determining whether grid transformers can be temporarily overloaded during peak demand without risking damage because the ambient temperature happens to be favorable that afternoon.

Georg started in 2007, back when nobody called it AI. Back then, it was just "data analytics" and "model building," using physics simulations and statistical methods to help power plant operators squeeze a few more months of life from expensive equipment. The goal was simply to determine whether you could run a turbine blade for another season rather than replace it, since the manual said it was due, saving millions while keeping the lights on.

Georg's physics background helps. He thinks in systems, in probabilities, in tolerances measured to fractions of a millimeter because in his world, manufacturing tolerances that seem insignificant, mere micrometers of deviation, can mean the difference between a component lasting 20 years or failing catastrophically after five.

"We have a lot of uncertainties and variations in all the data we are collecting," he explained to me during our interview. "Manufacturing tolerances that are very tiny, in operation, make a big difference in the durability of the parts."

This is a world where 99.9% accuracy leaves little margin for complacency. Where "pretty good" predictions are not enough in safety-critical industrial environments. Where experimental AI cannot rely on enthusiasm alone, but must demonstrate measurable reliability before it is ever deployed.

Yet Georg's team has been deploying increasingly sophisticated AI for nearly two decades, from early probabilistic models to computer vision systems monitoring factory processes to generative design algorithms optimizing turbine blade aerodynamics. Not because it's trendy. Instead, the complexity of managing modern energy systems with their mix of unpredictable renewables, real-time demand fluctuations, and equipment operating at the edge of physical limits has become impossible for humans to handle alone.

"The only way to manage that complexity," Georg told me, "is to use methods of artificial intelligence."

Here's what makes Georg's story fascinating: while headlines were dominated by multibillion-dollar funding rounds and sweeping restructuring across parts of the tech sector, the AI experts at Siemens Energy were working on a very different kind of challenge. Their systems had to prove themselves not through viral demos, but through sustained, real-world performance. They needed to operate continuously and reliably, supporting energy infrastructure entire societies depend on every day for years to come.

This is the AI we rarely talk about.

The AI that doesn't generate headlines or TikTok-style feeds. The AI that doesn't ask for our trust but earns it through years of reliable performance, gradually expanding its responsibilities as it proves worthy of greater authority. The AI that doesn't just ask for trust but earns it.

But here's what we're getting wrong: we're treating "AI" as if it were a single monolithic thing that we either trust or don't trust, when in reality we live with dozens of different AI systems that deserve radically different levels of trust. Some AI, like Georg's grid management systems, has been earning our trust for decades through consistent performance, and another AI, launched last week, is asking for faith it hasn't yet earned.

The fundamental challenge is that we haven't learned how to calibrate trust; in other words, we haven't learned to place our trust wisely based on what we're actually dealing with. We treat trust as a binary, on or off, when in reality it is contextual, conditional, and earned. We grant too much trust, too quickly, to some systems, while overlooking others that have quietly proven themselves over years of real-world performance.

Earned trust is the signal that systems have demonstrated reliability for a specific task, under defined conditions, with clear safeguards and human accountability. And, noise emerges when trust is granted prematurely: when scale, fluency, novelty, or hype substitute for evidence; when systems are trusted because they are impressive rather than because they are proven.

This chapter is about learning that distinction, how to separate signal from noise, so we can place trust wisely, based not on what AI *can* do in theory, but on what it has actually *earned the right* to do in practice.

The real risk is not artificial intelligence itself, but misplaced trust.

■■■

The Signal First: AI That Has Already Earned Our Trust

Let me show you what trustworthy AI actually looks like, because you've been trusting it for years without realizing it.

I start my day putting on music in my kids' rooms. So far, their music tastes are mine: Bach Radio and Latin Dance Mix. Behind these personalized playlists lies Spotify's recommendation engine, a sophisticated AI system that not only remembers what we played yesterday but also understands music itself. Right now, it's analyzing Glenn Gould's 1981 recording of Bach's Goldberg Variations: the tempo around 60 beats per minute, perfect for waking up gently; the solo-piano instrumentation; the key signature of G major, bright and uplifting; and mapping the emotional arc of each variation. When it suggests the next piece, it matches all these characteristics and also notices that millions of other listeners love this recording.[1]

For Latin Dance Mix, the algorithm weights different signals: strong percussion patterns, songs in specific key ranges that make people move, and tracks where the rhythm section dominates. It also reads the titles I give my playlists, scans what music journalists write online, and notices patterns in when and how I listen.

After 20 years of refinement and hundreds of millions of listeners, these systems have become eerily intuitive. My kids will grow up thinking machines simply understand what music should come next.

As I transition into my workday, I queue up podcasts: Hard Fork, Pivot, The Gray Area, The Ezra Klein Show, Your Undivided Attention. Here's where something different happens, or rather, doesn't happen. Apple Podcasts isn't pushing recommendations at me based on sophisticated AI. No "listeners like you also enjoyed…" Its Listen Now tab surfaces recommendations, but the interface centers the shows I've explicitly chosen to follow rather than continuously steering me toward adjacent content.

I prefer it this way. I know what podcasts I want. I've built this list through conversations with friends, articles that intrigued me, and episodes people forwarded me. I don't want a machine predicting that because I listen to tech policy discussions, I might enjoy true crime podcasts, even if statistically, many people who share my listening profile do. Sometimes AI's pattern-matching magic delights us. Sometimes we want our choices to be messier, more human, less optimized. Knowing the difference between where AI is working and where it isn't lets us decide when to ride the algorithm and when to chart our own course.

I take a Waymo to The Battery, my usual spot for meetings and interviews, located in the historic Jackson Square neighborhood. The car glides through the streets of San Francisco with an eerie confidence. Through the windshield, I watch what the car sees: pedestrians lighting up on its internal display, cyclists tagged and tracked, traffic lights recognized, other cars' trajectories predicted milliseconds into the future. Right now, multiple cameras, LiDAR sensors, and radars are generating a three-dimensional map of everything within 300 meters, which is refreshed continuously.[2] Computer vision algorithms process this data stream, machine learning models predict what pedestrians and other drivers might do next, real-time mapping systems understand construction zones and changing road conditions, and decision-making frameworks navigate complex urban environments. Each ride generates data that feeds back into the system, training future iterations. Waymo's logged over 100 million autonomous miles since 2009, learning from every awkward pause, every smooth merge, every edge case.[3] It's among the most extensively tested AI in the world. Yet, compared to the financial systems I'll use today, some of which have been running since the 1970s, it's still relatively new.

I love The Battery's chef and sommelier, who carefully curate menus and source produce from farm-to-table artisanal shops. In an urban environment like San Francisco, that's how a restaurant makes a statement. It has to be, because much of the world's produce is now mass-produced with the help of automated systems. AI has transformed modern agriculture in ways most consumers never see. Large-scale crops, wheat, corn, rice, soybeans, and vineyards increasingly rely on AI systems.[4] Drones fly over fields, their multispectral cameras feeding data to computer vision algorithms that detect plants stressed by too much sun or too little nitrogen, early signs of disease, and weeds among crops.[5] Machine learning models optimize irrigation schedules down to the hour using soil moisture sensors, weather predictions, and growth stages. Other algorithms predict yields months in advance. Automated harvesting robots powered by computer vision identify ripe produce and pick it without damage.[6] Supply chain optimization systems route produce to minimize spoilage.[7]

From seed to table, AI has fundamentally reshaped how food reaches us. The salad greens, the bread, the wine – are all increasingly touched by algorithms we rarely think about.

I have a tab running at The Battery, which knows my identity and links to my payment method. Everything is automatically settled when lunch or afternoon tea is done. My guests enjoy the seamless wrap-up, and I save a few minutes by not having to track down waiters, wait for bills, or sign receipts. Everything runs through my credit card, but what appears effortless is actually a complex web of AI-powered fraud detection working in the background.

What we don't see is the split-second interrogation happening as we speak. Algorithms that have been evolving since the early 1990s are asking: Does this transaction fit her regular pattern? Is she usually in San Francisco on Tuesday afternoons? Has her card been used anywhere else recently? Is this restaurant amount within her typical range, or is it a suspicious spike? Pattern recognition systems analyze transaction patterns in real time: flagging anomalies, assessing risk scores, verifying location data against my typical spending patterns, and using behavioral patterns to ensure the transaction is genuinely mine. Statistical models trained on billions of transactions continuously distinguish legitimate purchases from fraud, all within milliseconds between swipe and approval.[8] These fraud detection systems have processed literally trillions of transactions over decades, building up an almost supernatural sense for what's normal and what's not.[9]

When my card was first approved years ago, my creditworthiness was judged in seconds by scoring systems born in 1989, the same year Taylor Swift was born, the same year the Berlin Wall fell.[10] These statistical models have been refined over almost four decades of repaid and defaulted loans, learning and relearning what makes someone likely to repay. They're so embedded in the financial system that we no longer think of them as AI. They're just how credit works.

On days when Waymo or Uber takes forever because demand surges or there are routing issues, I take the bus. In San Francisco, many buses are electric trolley buses powered by overhead wires. Yes, San Francisco, a tech center, has more overhead utility lines and bus cables visible than underground infrastructure in many neighborhoods. This makes me wonder about something less visible yet equally complex: how AI manages the electrical grid as the mix of intermittent renewable energy sources increases in global consumption.

This brings us back to Georg's world. Right now, somewhere in California, solar panels are generating varying amounts of power depending on cloud cover. Wind turbines are spinning faster or slower with the breeze. Homes are turning on air conditioning as the afternoon heats up. Data centers are ramping up processing for the evening workload. And a vast network of AI systems is conducting this chaotic orchestra. Grid management AI systems must constantly balance supply and demand in real time and predict when solar panels will produce less due to cloud cover, when wind farms will ramp up or down, when households will draw peak power, and, critically, when massive data centers powering AI itself will spike their energy needs. These AI systems use weather forecasting, historical consumption patterns, and real-time sensor data to orchestrate a complex dance: routing power from where it's generated to where it's needed, storing excess in batteries, and even negotiating with large consumers to shift their usage to off-peak hours.[11] The grid is rebalanced every few seconds based on predictions made by machine learning models trained on millions of sensor readings. The irony is not lost on me: we need AI to manage the grid that powers AI itself. The system is recursively dependent on its own intelligence.

When I get home, I check the stock market once a day. It is a great way to stay attuned to perceptions of AI's value to the economy, geopolitical impacts worldwide, and shifting investor sentiment. But the market I'm checking has itself become an AI creature. High-frequency trading algorithms based on pattern recognition and predictive models operate in microseconds, far faster than any human could perceive, let alone react to.[12] They're reading news headlines and social media feeds in real time, detecting patterns in price movements, calculating probabilities, and making and closing positions before I could even finish reading a stock ticker.

When I check my portfolio, I see dozens of company tickers scrolling by. How many of these companies are using AI? How is AI changing their core competencies? And who manages all their administrative functions, from sales to marketing to commerce to data analytics? For most of them, I think of Salesforce.

What's remarkable about all this AI is that you probably didn't think of any of it as "AI" until I pointed it out. Spotify's recommendations feel like good curation. Waymo feels like an excellent driver. Credit card fraud detection feels like something you never think about at all, which is precisely the point. The electrical grid feels like electricity just works. The stock market feels like a market.

This AI has earned our trust not through marketing campaigns or impressive demos, but through something far more valuable: consistent performance over time. These systems prove their value not by being remarkable in controlled environments, but by working reliably day after day, year after year, handling edge cases gracefully, failing safely when they do fail, and gradually expanding their capabilities as they prove themselves worthy of greater responsibility.

The AI we trust most is often the AI we notice least. This is because reliable technology becomes invisible. When a system works so consistently that we stop thinking about it, that's not a failure of transparency. That's the highest compliment you can pay to technology.

How AI Became Invisible Infrastructure

The popular story suggests that artificial intelligence arrived suddenly in November 2022 – when ChatGPT appeared, 100 million people signed up in 60 days, and AI seemed to materialize everywhere at once. That moment was real. But it wasn't the beginning of AI. It was the moment AI became visible.

What we call "AI" today is not a single invention. It is an accumulation.

For more than 70 years, different kinds of AI have been built to solve different problems: rule-based systems for logic and control, statistical prediction models, machine learning for pattern recognition, deep learning for perception, and, more recently, generative and agentic systems for language and action. Crucially, these approaches did not replace one another. They stacked.

Each generation absorbed the previous one's lessons, extended its reach, and embedded itself more deeply into how organizations operate. Over time, many of these systems stopped feeling like "AI" at all. They became infrastructure, quiet, dependable, and largely invisible, just like how Georg has described.

Trust was formed through systems that handled edge cases, failed safely, improved incrementally, and, over the years, proved they could be relied on primarily for specific tasks under specific conditions.

AI becomes infrastructure, and we no longer experience it as technology, but as life continues uninterrupted. Try, for a moment, to live a single day without AI, not by turning off your phone or declaring a digital detox, but by simply paying attention.

■■■

The Noise: When Experimental AI Demands Immediate Trust

Now that you understand what trustworthy AI looks like, systems that have earned trust through decades of consistent performance, let's contrast that with the AI dominating headlines today.

At first glance, nothing about today's AI feels invisible. It is everywhere. It writes, speaks, generates images and video, and increasingly mediates how we search, decide, and create. But what defines this moment is not visibility. It is misdirection. What is most visible is not what matters most.

It's easy to focus all our attention on the newest and hottest generative and agentic AI. But these new generations of AI have yet to be tested by time, filtered through failure, or proven in the domains where they truly belong. They remain in what we discussed in Chapter 2 as *elite wonder*, where massive investment pours into inception, testing, creation, and early adoption.

OpenAI, Google, Meta, Anthropic, Musk's Grok, and China's DeepSeek have dominated headlines in newspapers, magazines, and online platforms. They are the companies that create the so-called foundational models, which are general-purpose engines that can summarize, analyze, reason, write, code, create images or video by understanding patterns in language, vision, and logic. These foundational models power the most visible part of artificial intelligence we see today. What remains largely unseen is how they fail, where they break, and who absorbs the cost when they do.

The migration from traditional search to AI chatbots signals the first visible shift in how we access information. By October 2025, ChatGPT alone reached 800 million weekly active users.[13] Half of US adults now use large language models such as ChatGPT, Gemini, Claude, and Copilot.[14] These numbers create a sense of social proof. But they reveal little about trustworthiness – how often outputs are wrong, biased, incomplete, or contextually misfit.

Some of these AI companies have drifted into social media territory, building platforms designed less for utility and more for attention-grabbing and data collection.

In late September 2025, within days of each other, Meta launched an AI video platform, Vibes, and OpenAI launched Sora, an AI-native social network in which every video was AI-generated.[15] Both had immediate challenges. In Sora the Cameos feature lets users upload biometric data to insert themselves into any scenario imaginable.[16]

Within days, the platform was flooded with deepfakes of public figures in compromising scenarios and copyrighted characters.[17] Vibes was quickly panned by critics and users alike as little more than "AI slop," a term that's come to describe the flood of low-effort, synthetic content clogging feeds with noise rather than value.[18]

These systems generated massive attention precisely because they were engineered to do so. They were optimized for engagement, growth, and cultural saturation. Success was measured in active users, engagement curves, and valuation multiples that the product features contribute to. But attention is not durability. Six months later, OpenAI shut Sora down as the audience had evaporated and the app was burning $1 million a day.[19]

In enterprise settings, we have seen companies announce productivity ambitions while restructuring labor. Microsoft cut 15,000 employees in 2025 while investing heavily in AI.[20] Amazon's CEO, Andy Jassy, stated that fewer people would be needed to do many existing jobs.[21] Meta eliminated thousands of roles while publicly framing AI as a mid-level engineer.[22]

Beneath the noise are the organizational experiments that struggle, the workflows that stall, and the human roles that are gradually reconfigured before replacement systems are fully ready. In most cases, this is less about recklessness, more about what modernization looks like under uncertainty. Leaders are moving in good faith, under pressure to act quickly, with limited precedent and incomplete information. The risk is that leaders reconfigure decision-making before they have reliable ways to know where human oversight still matters most.

When AI appears ubiquitous, when peers adopt it, when institutions frame it as unavoidable, trust becomes ambient rather than deliberate. The question quietly shifts from *Is this system ready?* to *Why would I opt out?* This is why the noise matters. It crowds out judgment.

Yet these systems are still early in their trust cycle. They are moving from elite wonder toward mass exposure before their failure modes are fully understood. They are asking for trust not because they have earned it, but because they are already everywhere.

The stated inevitability of the AI boom, amplified by dazzling headlines and constant product releases, compels acceptance. AI starts to feel less like a choice and more like the weather: something happening to us rather than something we are deciding.

The contrast could not be starker. While teams like Georg's spend years validating each system before deployment, building redundancy and expanding scope only as reliability is proven, frontier AI companies release experimental systems to hundreds of millions based on impressive demos and future promises. Learning to see what is absent, not just what is impressive, is now a core skill of the AI era.

■■■

Calibrating Trust: Three Models of How AI Earns Institutional Trust

Understanding how some AI systems earned our trust while others remain experimental helps us calibrate our expectations for frontier AI, learn from the AI embedded in our daily life, and institutionalize trust. Let me show you three companies that have built AI systems the world depends on, where trust wasn't demanded through hype but earned through systematic reliability over time.

But first, a point that's easy to miss. Most of the AI shaping your daily life never comes directly from an AI lab to you. It moves through a long supply chain: from model builders to platforms to enterprises, and only then to consumers. By the time you experience a stable power grid, faster fraud detection, or smoother customer service, the AI has already been reshaped by multiple layers of human and organizational decision-making. Trust isn't what you feel at the point of use. It is what organizations build, or fail to build, long before AI ever reaches you.

Three different paths to the same destination: AI so trustworthy it becomes invisible. Siemens Energy achieves invisibility through operational excellence that consumers rarely notice. Mastercard achieves invisibility through brand trust, enabling innovation to happen behind the scenes. Salesforce achieves invisibility by making trust itself so visible and disciplined that AI can safely operate in the background.

Different approaches. Same outcome. AI that works so reliably you forget it's there.

Siemens Energy: Trust Through Absence of Failure

Georg Rollmann's story exemplifies what I call the "trust through absence of failure" model. This is AI at its purest form of infrastructure – working so well you never think about it, managed by people you'll never know unless something goes catastrophically wrong.

Georg and his team work on AI systems that help manage critical infrastructure across global energy networks. Gas turbines generate electricity for entire cities. Wind farms feed gigawatts into national grids. Predictive maintenance systems are monitoring thousands of power plants for early signs of failure. Manufacturing quality control across 90+ factory sites. Design optimization systems that create more efficient turbine blades. And yes, grid management systems balance supply and demand in real time across entire countries. The AI here is essential to how modern energy systems function.

Consider what's happening in the world's energy systems right now. In 2025, China added more than 430 GW of new wind and solar capacity, lifting total installed renewable capacity above 1,800 GW and pushing renewables to more than 60% of the country's total capacity.[23] The EU added 65 GW of solar in 2025, bringing total installed solar capacity to 406 GW and hitting its mid-decade target.[24] In the United States, the first nine months of 2025 saw more than 30 GW of solar and 12.9 GW of battery storage come online, accounting for 85% of all new power capacity added to the grid.[25]

All these regions share a fundamental challenge: unlike coal or gas plants, which run continuously at a predictable output, solar and wind are intermittent and weather-dependent. The grid must constantly balance fluctuating supply and demand, second by second, to maintain stable voltage and frequency, keeping your lights on and preventing equipment damage. Humans cannot physically do this at the required scale and speed. AI can.

Siemens Energy's approach spans what Georg calls an "AI capability model," five distinct but interconnected domains that reflect the company's engineering culture. First is advanced machine learning for computer vision, pattern recognition in time series, and the construction of surrogate models for design optimization. Second is probabilistic modeling and uncertainty quantification, critical for systems where manufacturing tolerances measured in micrometers determine whether equipment lasts five years or twenty. Third is time-series analytics for the massive streams of IoT data from thousands of machines in the field and 90+ manufacturing sites globally. Fourth is generative AI, which they've been using far longer than the current hype cycle – for engineering design optimization, generating new turbine blade geometries that burn less fuel per megawatt produced. Fifth is optimization methods for multi-objective, stochastic problems, such as optimizing customer bids or designing new components.

This taxonomy is operational. When Georg's team evaluates a potential AI application, they assess which capability domains apply, what validation protocols are required, and how to integrate with existing domain expertise. "You don't take an off-the-shelf large language model to steer your power plant," Georg explained matter-of-factly. "This is just not going to work, because no domain knowledge is included."

The trust dynamic in energy is fundamentally different from consumer-facing AI. You don't opt into or out of the AI managing your local power grid. You simply expect the lights to work when you flip the switch. This creates trust through the absence of failure. AI earns trust not by being visible or celebrated, but by being so reliable you never have to think about it.

Siemens Energy's predictive maintenance AI analyzes sensor data from turbines to predict component failures weeks or months in advance, enabling scheduled maintenance rather than catastrophic breakdowns. When this works, which is nearly always, nobody notices. Your electricity simply never goes out. The wind farm keeps spinning. The grid stays balanced. Georg and his team remain anonymous.

When it fails, even briefly, everyone notices immediately. Hospitals lose power. Traffic lights go dark. The operational leaders become very visible, very quickly.

This is AI operating under completely different success criteria than a consumer chatbot. For ChatGPT, 95% accuracy might be acceptable; if it hallucinates occasionally, it's annoying but not catastrophic. For grid management AI, 99.9% uptime isn't an achievement worthy of celebration, and it's the minimum baseline expectation.

Georg and his team aren't trying to make their AI visible to end consumers. That would be counterproductive. If you're thinking about how the grid works, something has gone wrong. Instead, they focus on transparency and accountability to the entities that do need to understand the systems: grid operators, regulatory bodies, energy ministries, and safety inspectors.

When a major storm threatens the grid, Georg's AI can optimize load shedding to minimize outages, calculating in milliseconds which areas to prioritize based on hospital locations, population density, and grid topology. But humans still make final decisions about which neighborhoods to leave in the dark first, because those decisions involve social and political tradeoffs that can't be reduced to algorithms. The AI provides superhuman analysis speed; humans provide judgment about values and edge cases. This division of labor – AI for speed and pattern recognition, humans for ethics and unusual situations – represents decades of refinement in how industrial systems integrate machine intelligence.

The validation process reflects these stakes. Instead of replacing existing processes, new AI systems augment them, running in parallel for months while operators compare predictions against reality. "It's installed, and then for some months, it runs with the normal process," Georg explained. "And then people see, ah, it's helpful. I was right there, and I made a mistake. I can correct that. And then at some point, it's more and more being accepted."

They always have fallback solutions. In engineering, they can try the new AI method and revert to old approaches if it doesn't work. In manufacturing, they can install a new system and simply switch it off if it doesn't help. In service and predictive maintenance, if the AI generates more alerts than before, operators review them. False alarms are preferable to missed failures because they always keep the standard system running in parallel.

Consumers care about whether electricity is available and accessible. Institutional trust builds on infrastructure reliability. Regulatory oversight, professional expertise, technical redundancy, and safety protocols ensure that. The AI managing the grid behind the scenes has been working reliably for years, tested by millions of hours of operation, refined through every storm, heatwave, and equipment failure. It's earned trust through consistent performance, even if most beneficiaries don't know it exists.

"If AI is related to safety, nobody would accept playing around with that," Georg said flatly. "That is, I think, totally clear."

This is the operational leader's model: succeed by staying invisible. That's the highest compliment you can pay to critical infrastructure. It's so reliable you forget it's there.

Mastercard: Trust Through Brand That Buys Time for Innovation

If Siemens Energy shows how invisible operations build institutional trust, Mastercard demonstrates the opposite: how visible brand trust enables invisible innovation. The consumer-facing brand provides permission to take technical risks behind the scenes.

I first spoke with Raja Rajamannar, Mastercard's Chief Marketing and Communications Officer, six years ago when I launched the Safety Matters Summit. Even then, before the generative AI explosion, Raja demonstrated remarkable foresight about how safety forms the baseline for innovation, and how innovation success reinforces brand trust. Fast forward to October 2025, and that foresight has proven prophetic.

You probably don't think about AI when you tap your card at a coffee shop. You should. In the fractions of a second between your tap and the approval beep, Mastercard's AI systems analyze signals from your transaction, compare them against decades of fraud patterns across more than one trillion data points, and make a risk determination that protects both you and the merchant.[26]

When I asked Raja how Mastercard maintains brand trust while rapidly deploying technologies that introduce genuine uncertainty, he drew me a framework that explains their entire strategy:

"If I look at you and ask, is Tiffany trustworthy?" Raja began. "First and foremost are your values, your integrity, your ethics. You are trustworthy because there is a fundamental relationship based on a foundation of trust. That's the first level. The second order of trust is capabilities. Tiffany has been leading AI startups into significant growth. Tiffany has founded a non-profit focused on safety. Tiffany is well-versed in AI and can explain it in relatable ways. She has subject matter expertise and experience. Do I believe that? That builds the second-order trust. The third level is adjacencies. Tomorrow, if Tiffany tells me she has a special solution that will make my hair grow back, do I believe her? That belief is not based on foundational trust or domain trust. Does she have expertise in this domain, which I did not associate with her till now?"

This pyramid explains Mastercard's entire approach: build visible brand trust that buys you time to develop invisible technical capabilities, which then enable you to expand into adjacent domains where you have no natural right to play.

The foundational trust comes from what Raja calls "doing something good for society, not directly for the company." It's strategic trust-building at scale. Raja told me that through its partnership with the World Food Programme, Mastercard has helped deliver the equivalent of 160 million school meals to children in underprivileged communities since 2012.[27] They partnered with Stand Up to Cancer and raised over $75 million. When I asked how this works, Raja explained: "We create a small movement where, instead of using cash or any other card, you just use your Mastercard. It's the same price. But I, as Mastercard, will take a small percentage of my profit and give it to Stand Up to Cancer. Then, at MLB's All-Stars and World Series, we do what we call a Stand Up to Cancer moment, where we ask people to stand up with a placard that says, "Who am I standing up for… my mom, my sister, whatever it is." It's a very emotional moment. And Mastercard is there subtly throughout. We're not bragging, we are just enabling and facilitating."

Raja emphasized that when Mastercard commits to a cause aligned with its values, it stays committed for the long term, not as a flavor-of-the-month effort, but as an enduring part of the company's identity. That sustained commitment is what builds trust.

This brand-building bought something priceless: time to build technological capabilities that consumers never see. While Mastercard was publicly known for feeding children and funding cancer research, they were steadily acquiring and integrating data and AI companies. They acquired Brighterion, a company specializing in AI-driven fraud prevention.[28] They added NuData's behavioral biometrics, which monitors hundreds of subtle signals such as keystroke dynamics, typing speed, and how you hold your device to distinguish legitimate customers from impostors.[29] They acquired Finicity for open banking capabilities,[30] Dynamic Yield for personalization,[31] and are now deploying generative AI that can boost fraud detection rates by up to 300% while reducing false positives by more than 85%.[32]

Each generation of AI built credibility for the next. But brand trust does not only buy time, it also imposes discipline. Mastercard builds its AI Governance Council, consisting of its Chief Privacy Officer, Chief Data Officer, Chief Security Officer, and the EVP of AI to oversee all AI activities against the company's data responsibility principles. It has insisted that acquisitions meet its Privacy by Design standards, even when that meant slowing integration timelines considerably. In at least two cases, the framework led Mastercard to walk away from acquisitions entirely.[33] The same trust that gives a company permission to innovate demands that it uphold the principles that earned that trust in the first place.

Now consider the newest leap: agentic payments. AI shopping assistants that can buy things on your behalf sound terrifying. It's like you're putting a piece of code in charge of your money. But Raja reframes it brilliantly: consumers already do this. "Amazon auto-renewal, that is an agent. Maybe [it's] not run by Agentic AI the way we've been defining it, but I'm delegating the decision to it: every month you can keep debiting my credit card. We have already crossed that big inertia point. We should not confuse [consumers] or complicate life [for consumers] to say this is agentic AI. We need to showcase to them, in very simple terms, how it makes life easy for them, and how it is safe because it is backed by Mastercard." Mastercard launched Agent Pay with agentic tokens: cryptographic credentials tied to specific AI agents, merchants, and spending parameters you define.[34]

This is the Mastercard model: visible brand trust buys time and permission for invisible technological innovation, which reinforces brand trust, which enables expansion into new domains. It's a flywheel that takes years to spin up but becomes nearly unstoppable once it's moving.

Salesforce: Trust Through Organizing Principle

If Siemens Energy demonstrates invisible operations that build institutional trust, and Mastercard shows visible brand trust that enables innovation, Salesforce represents a deliberate strategic choice: making trust itself the visible, unifying principle of an entire enterprise.

Here, trust is the organizing mission that shapes every product decision, every acquisition, every line of code.

I know this firsthand. I worked at Salesforce, and one of the most profound lessons I learned there was how trust functions as a constant alignment mechanism. In a company that has grown from a single sales software product to a sprawling empire of clouds, Sales, Service, Marketing, Commerce, Data, Analytics, spanning every enterprise function imaginable, trust is the thread that keeps everything coherent as Salesforce's #1 value.

When you're building software that manages customer relationships for 150,000+ companies, including the majority of Fortune 500 firms, and any one of your AI features cascades across billions of customer interactions, trust can't be an afterthought.[35] It has to be the foundation.

Marc Benioff, Salesforce's founder and CEO, built the entire company on a radical premise: convince businesses to put their most valuable data – customer relationships, sales pipelines, service interactions – not on servers they control, but on "the cloud," managed by someone else. In 1999, when Salesforce launched, this was a preposterous ask. The customer roster grew from small and medium-sized businesses to major enterprises – companies that had spent decades building on-premise data centers with armed guards and complete physical control. Marc was asking them to trust his company with everything.

The trust notion Salesforce built to overcome this skepticism evolved in phases, each responding to the technological and competitive landscape of its time. In the early years, when Marc was trying to convince skeptical CIOs to put data on the cloud, trust meant three things: security (your data won't be stolen), reliability (our systems won't go down), and performance (it'll actually be faster than your on-premise servers). These were existential requirements. If Salesforce failed on any dimension, the entire business model would collapse. So, they invested heavily in redundant data centers and enterprise-grade security. They pioneered the concept of a "trust site," a public dashboard showing system performance, scheduled maintenance, and any incidents.[36] Such transparency, showing when things went wrong, not just when they went right, built credibility with IT leaders who were used to vendors hiding problems.

As the company grew and acquired more capabilities, trust evolved. The launch of AppExchange, Salesforce's app marketplace, introduced a new risk: Salesforce was asking customers to install third-party applications directly into their business systems. If a partner app had a vulnerability, it could threaten the integrity of the shared infrastructure. Salesforce responded by building a mandatory Security Review, not a simple functionality check, but a strict audit where Salesforce engineers performed penetration testing on partner code.[37] This process allowed a Fortune 500 company to install an app from a small startup, knowing it met the same security benchmarks as Salesforce itself. When Salesforce launched Marketing Cloud through the ExactTarget acquisition, trust expanded to include permission-based marketing and anti-spam controls. When they added Commerce Cloud via the Demandware acquisition, trust meant PCI compliance and secure payment processing. For customers in highly regulated sectors such as finance, healthcare, and government, Salesforce developed Shield, a suite of compliance tools designed to satisfy strict legal requirements. They built the Hyperforce platform to solve critical global compliance issues by allowing customers to specify the geographic locations where their data would be stored. Each new cloud brought new trust requirements and shaped new trust dimensions.

Then came generative AI.

Einstein GPT in 2023 and Agentforce in 2024 introduced systems that not only analyzed data but also acted on it, including drafting emails, resolving tickets, and updating records. Trust now meant something new: preventing hallucinations, containing data leakage, ensuring explainability, and preserving human oversight.

Salesforce's approach was to build what they call the Einstein Trust Layer, wrapping all their AI in comprehensive safety systems: data masking that protects sensitive information before AI sees it, toxicity detection that prevents offensive responses, grounding that ensures AI responses are based on actual CRM data rather than hallu cinations, and audit trails that track every AI action.[38] This effort was required because AI was unproven at that level of autonomy.

But here's what makes the Salesforce model instructive: trust is embedded not just in the technology, but in the company's visible culture and communication. Every product launch emphasizes trust first. Every employee has it ingrained in them that "Trust is our #1 Value." When I was there, you couldn't escape it. Trust was in every employee's annual goals, referenced in meetings, and built into performance reviews.

This visibility serves a purpose. In an enterprise software company where most customers never meet the engineers building the product, where the technology is complex and the stakes are high, making trust visible and constantly discussing it ensures that everyone, from new graduates to senior executives, understands it's not negotiable. The consistency of this message over 20+ years is what allows Salesforce to rally teams around a coherent mission even as the company grows to nearly 90,000 employees and the product portfolio expands to encompass everything from sales to marketing to commerce to analytics to AI agents.

The result is that Salesforce has become a trusted infrastructure for enterprise AI, and a platform where companies can adopt and customize powerful AI capabilities without having to build all the governance, security, and oversight mechanisms themselves.

■■■

The Invisibility: Trust Is Earned, Not Claimed

Siemens Energy's operational excellence, Mastercard's brand reputation, and Salesforce's values-driven leadership each highlight a different entry point toward the same destination: AI is so trustworthy that it becomes invisible.

These three principles cover what enterprises must do to earn trust in AI:

Operations must prove reliability through consistent performance over time. Brands must build permission through values that buy time for innovation. Systems must embed discipline through visible governance.

For consumers, these become your signals: Has this AI proven itself over the years? Does the brand stand for something beyond profit? Is trust governance visible and serious? Are the organizations behind the AI investing in trust governance, or just in speed?

Six patterns emerge from these exemplary enterprise deployments. They distinguish AI that has earned trust from AI that merely demands trust.

Time validates. Georg's grid management AI has refined itself through millions of operating hours across thousands of power plants, learning from every storm and equipment failure over decades. Frontier AI? ChatGPT launched two years ago and is already deployed to hundreds of millions of users, asking for immediate trust based on potential rather than proven performance.

Stakes determine requirements. When Spotify misreads your music taste, you skip a song. When grid management AI fails, cities go dark. Georg was clear: "When AI affects safety-critical systems, no one will tolerate experimentation." Yet frontier AI is often deployed at a global scale, even as its failure modes are still being discovered.

Transparency serves stakeholders who need it. Salesforce makes trust radically visible by showing performance, outages, and everything else, because its entire business model is built on trust. Frontier AI, by contrast, often remains opaque, even to the engineers who train it. Part of this opacity is architectural: transformer-based models are less interpretable. But the larger gap is operational. A robust trust layer, one that safeguards systems and visualizes successes, failures, and degradation in real time, has yet to emerge for frontier AI.

Domain expertise is the difference between promise and production. Mastercard spent three years integrating Brighterion's fraud-detection AI to meet privacy and regulatory thresholds. General-purpose AI will increasingly demonstrate domain-level capability, but to run critical systems, deep integration, rigorous validation, and hard-won institutional knowledge are indispensable.

Fallback mechanisms are required. When Georg deploys new AI, it runs in parallel with existing systems for months. If it fails, immediate fallback procedures are in place. Frontier AI tends to move faster, straight into production, substituting for human judgment, often without the same margin for graceful failure.

Proof beats potential, always. When Salesforce publishes real-time system performance for 150,000+ companies on its trust site, showing every outage and every recovery, that's proof. When OpenAI raises billions based on what GPT-5 might achieve, that's potential. Proof is what you show to customers whose business depends on you.

Trusting Wisely

Trust, in the age of AI, is a skill. A skill we must learn, practice, and apply deliberately because not all AI deserves the same level of trust. That skill begins with discernment.

Signal is the AI that earns trust over time through evidence, limits, and accountability. Noise is AI that asks for trust before it has earned the right to be trusted. The challenge before us is learning how to tell them apart.

Across industries, trustworthy AI already exists, not in demos or declarations, but embedded in systems that quietly work. When your lights stay on, grid AI is balancing supply and demand. When your card clears, fraud models that have been refined over decades protect you. This AI did not persuade its way into trust. It earned it by becoming boring: reliable, constrained, governed, and proven over time.

Frontier AI represents something different. Not malicious. Not meaningless. But premature. It often asks for trust based on promise rather than proof, speed rather than discipline, and potential rather than performance. It reaches a global scale while its failure modes are still being discovered.

Preserving our humanity and our agency does not mean rejecting AI. It also does not mean surrendering to it. It means insisting that trust follows evidence.

Being wise about AI means recognizing when trust has been earned and when it is being borrowed. It means distinguishing systems built on operations, brand permission, people, and governance from systems powered primarily by hype.

The question, then, is not whether to trust AI but whether we are capable of trusting it wisely.

This distinction matters because the next systems will not only answer questions, but will also act. They will recommend, negotiate, decide, and move resources. And when AI acts on our behalf, misplaced trust stops being embarrassment and starts being harm.

Now that we know how to separate signals from noise, we can examine what happens when that skill fails, how control is lost, and why the discipline of trust is not optional.

So we turn to the risk map, not to frighten ourselves into rejection, but to learn the discipline that lets us accept AI's power without handing over our humanity.

Chapter 6

The Defeatable: Set the Boundaries

If I tried to list every risk associated with artificial intelligence, this chapter would never end.

Researchers at MIT cataloged more than 1,700 distinct AI risks across 74 frameworks.[1] Seventeen hundred! And this database keeps growing. I find the MIT AI Risk Repository standing as the most comprehensive effort to map what can go wrong: discrimination, toxicity, privacy violations, security vulnerabilities, misinformation, malicious use, human-computer interaction failures, socioeconomic disruption, environmental costs, and AI system safety failures... 74 frameworks are trying to make sense of it all.

Then there is the EU AI Act, organizing risks into tiers from "unacceptable" to "minimal."[2] NIST's AI Risk Management Framework has seven trustworthy characteristics.[3] The International AI Safety Report, authored by 100 experts from 30 countries, identifies "malicious use, system malfunctions, and systemic risks."[4] AIR 2024's four-tier taxonomy.[5] Anthropic's AI Safety Levels.[6] OpenAI's Preparedness Framework.[7] DeepMind's Critical Capability Levels.[8]

We have the maps. God, do we have the maps.

And yet when my neighbor asks me, "Should I let my kid use ChatGPT for homework?" or "Is that AI-powered security camera actually secure?" or "How do I know if this medical advice came from a bot?" All that cataloging evaporates.

Because the frameworks were not built for humans to make actual decisions every day, they were built for researchers categorizing phenomena, policymakers drafting regulations, and companies demonstrating compliance. They are essential for that work. If you want a comprehensive understanding of AI risk, read them. I mean that sincerely. The MIT repository and frameworks like NIST's are invaluable resources.

But this chapter has a different purpose.

I want to give you – whether you are deciding which apps to trust, which systems to deploy at work, or which policies to support as a citizen – a way of thinking about AI risk that maps to how you actually make trust decisions in daily life.

Three categories. Not seventeen hundred risks. Three patterns that tell you what kind of danger you are looking at and what kind of boundary will actually protect what matters.

This is applied sensemaking rather than academic completeness. I will let the databases catalog everything. This chapter guides us to recognize the pattern so we know how to respond to this inevitable wave that often invisibly impacts our lives.

Why Three Categories When We Have Seventeen Hundred Risks

In the first chapter, we established that trust, any trust, whether in people or systems, rests on four pillars: Truth, Competence, Benevolence, and Contextual Fit. When engineers and policymakers tried to translate these into measurable terms for AI systems, they created technical property frameworks with terms such as "validity," "fairness," "security," and "explainability."

Here is what I learned over years of building digital trust and user safety systems: when trust breaks, it does not announce which property failed.

A voice-cloned scam steals your grandmother's savings. Which property failed? Security, because authentication breached? Validity, because the system couldn't detect the fake? Explainability, because the victim didn't understand how to verify? All of them. Simultaneously.

Real AI risks do not respect the boundaries we draw around technical properties. They emerge at intersections. Professionals absolutely need these technical properties as they provide frameworks for building and auditing systems. AI engineers, security teams, ethics officers, and legal counsel: they battle through those properties every day, and that work is critical.

But when we are trying to decide – as a parent, a professional making a deployment choice, an executive weighing an investment, a citizen evaluating policy – whether something is safe enough for specific contexts, we need a different starting question. Not a replacement for technical rigor, but a human question that helps us know which properties matter most for our decision.

Not "which properties are we optimizing for?" but "what kind of danger am I looking at?"

That realization hit me at CHAI's annual workshop at Asilomar in the summer of 2025, six weeks after my daughter Cali was born. I was exhausted in that way only new parents understand: physically depleted, emotionally porous, newly aware of what vulnerability really means. I sat in rooms with AI researchers debating alignment failures and emergent capabilities. I thought of conversations I had had: journalists who warned that deepfakes would overwhelm verification, the security experts who tracked attacks scaling faster than defenses, and global policymakers who showed different value systems and contextualized regulations for their people.

Everyone was right. And yet, often, we tend to talk past each other.

Then it clicked. AI risks do not primarily differ by which technical property fails or which domain they affect. They differ by where the failure originates.

Some risks originate in us. AI is inheriting and systematizing the biases, blind spots, and structural inequities we never fixed before it arrived. They are old human ones, now rendered at machine scale.

Some risks originate from adversaries. AI removes friction from harm creation faster than we can defend against it, allowing malicious actors to weaponize capabilities that were once too slow, too expensive, or too technically complex to deploy at scale.

Some risks originate in the systems themselves. Optimization engines generate behaviors no one explicitly programmed and no one can fully predict, because they pursue objectives that outgrow our mental models and operational control.

I call these mirror risks, weapon risks, and ghost risks. For sure, I love categories, but it is more because I needed something that maps to how humans already understand danger. We already know the difference between a problem we created, a threat someone is using against us, and something behaving in ways we don't understand. We have different instincts for each. Different responses. Different boundaries we set.

Mirror risks demand that we fix upstream. We cannot automate what we have not made just. Weapon risks demand that we match speed or change the rules. This requires us to build defenses that operate at machine tempo. And ghost risks demand that we accept limits or seek alternatives, so we can build or adopt technology with restraint when we cannot predict or control.

Each category shows us what to worry about and how to set boundaries. Let us navigate this myriad of risks together.

■■■

Mirror Risks: When AI Reflects What We Failed to Fix

Mirror risks arise when AI inherits failures that existed long before it did and systematizes them.

A biased recruiter might interview 10 candidates inconsistently, sometimes letting prejudice influence decisions, sometimes catching themselves, and sometimes getting called out by a colleague. An AI trained on that recruiter's decisions could review 10,000 candidates with perfect, mechanical consistency. Every. Single. Time.

The mirror shows us our failures without the noise of human variation. Without the plausible deniability. Without the exceptions that let us pretend the problem was the work of individual bad actors rather than systemic rot.

That is why mirror risks are the most uncomfortable. They implicate us.

The Bias Nobody Wanted to Systematize

In 2014, Amazon built an AI recruiting tool that seemed like the future: automatically review applications and identify the best candidates. By 2015, they realized it systematically downgraded resumes that included "women's" as in "women's chess club captain." It penalized graduates of all-women's colleges.[9]

The model had not been programmed to discriminate. It learned from a decade of hiring data that reflects a male-dominated recruiting process.

Amazon's story became famous. But the pattern is everywhere. A widely used algorithm in US hospitals systematically underestimated Black patients' health needs because it used healthcare spending as a proxy for need, embedding decades of unequal access into automated decisions about who receives care.[10]

In 2024, researchers at the University of Washington tested major language models on hiring scenarios. The results were stark: LLMs favored White-associated names 85% of the time. More striking: they never favored Black male names over White male names. Not once. Not in millions of simulated decisions.[11]

This is AI inheriting discrimination from training data, from the world the data reflects, from patterns so deeply embedded we stopped seeing them as choices.

Niki Parmar knows this tension intimately. She co-authored "Attention Is All You Need," the 2017 paper that introduced the transformer architecture, the very foundation for ChatGPT, Claude, and nearly every major AI system today.[12]

"The thing is that if the data or the intent is bad, [the model is] always going to pick up on that and learn that," she told me. "So sometimes we need to actually counteract the biases that naturally exist in the world."

She paused, then continued: "Why shouldn't we? Because maybe it can help us also correct the mistakes that we as a society [have made]."

The architecture is fundamentally agnostic. It learns patterns from data. If the data reflects historical injustice, the model will systematize that injustice unless we actively intervene. This is not a technical problem. It is a governance problem.

Amazon's solution was to abandon automated screening for roles requiring human judgment about potential. They recognized that no model trained on biased outcomes would produce unbiased decisions, no matter how sophisticated the architecture.

But here is what makes this a mirror risk: removing AI would not eliminate hiring discrimination. It would make it slower, less systematic, and more deniable. The discrimination was there first. AI made it undeniable.

The boundary mirror risks demand is to fix discriminatory practices that created biased data, confront structural inequalities, and maintain human judgment where automation quietly replaced it. Governance needs to go before optimization. We cannot automate what we have not made just.

Platform Toxicity and What Algorithms Learned to Amplify

During my years at Spectrum Labs and OpenWeb, we built systems protecting over a billion users from online harm. We deployed classifiers, which are models that learn from examples and then decide which predefined category a new piece of data belongs to, to detect hate speech, harassment, and coordinated manipulation. Abuse dropped. Then adversaries adapted: using coded language, creative misspellings, images instead of text, and migrating to under-moderated platforms. We updated the models. They adapted again.

The underlying problem was not the tools people used. It was human intention combined with platform incentives that rewarded engagement over well-being.

Social media companies built systems that algorithmically amplified outrage because outrage drove engagement, which in turn drove advertising revenue. The platforms were definitely not neutral town squares. These were recommendation engines optimized for time-on-site, and time-on-site correlated with emotional arousal, to be clear, especially anger and fear.

AI systematized online cruelty. The same toxic incentives that once produced sporadic harassment now produce consistent and measurable industrialized abuse.

The boundary mirror calls for reform of incentive structures that reward engagement over well-being, updates to consent frameworks for data collection, and platform accountability for algorithmic amplification. These are generation-spanning problems requiring regulatory pressure, business model innovation, and cultural change. We cannot outrun what the mirror reflects. We have to repair it first.

The Information Collapse That Preceded AI

Kevin Goldman spent decades inside American journalism, spanning *The Wall Street Journal*, *The New York Times*, the *Philadelphia Inquirer*, *Variety*, and *Newsday*. He has the institutional memory most people lack.

He asked me to watch the 1987 CBS Evening News Christmas card on YouTube. In it, I saw nearly 200 faces. Foreign correspondents from Beijing, Johannesburg, the Persian Gulf, Rome, and Tokyo. Producers, editors, and camera operators. Bureau chiefs across five continents. A distributed system designed for verification. I understood why his voice carried something more than nostalgia.

"From then till now…" Kevin said, and paused.

What happened between 1985 and today was a systematic erosion of trust infrastructure.

Jennie Baird, who retired after chairing product organizations across News Corp and the BBC, did not start our conversation about AI risk with deepfakes. She began with incentives.

"Facebook and Google basically captured distribution. And with distribution went the advertising revenue that had funded journalism," she said.

The pattern was predictable. In 2013, Facebook boosted news content, driving massive traffic. Publishers invested heavily in Facebook-optimized content. Then in 2018, Facebook pivoted away from news. Traffic collapsed. Publishers discovered they had built on rented land, and the lease was up.

"Over the past 10 years, journalism has shifted toward either clickbait or content that plays to extremes, because that's what the algorithms like," Jennie told me.

The economic structure inverted before ChatGPT existed. Platforms optimized for engagement over truth. Business models could not sustain verification. Institutional trust hit historic lows. By the time AI arrived, capable of generating convincing but false content at scale, the information ecosystem was already failing.

The Atlantic Council's Digital Forensic Research Lab documented how pro-Kremlin networks weaponized the very foundation of crowdsourced knowledge: they flooded the information ecosystem with polluted sources that well-meaning Wikipedia editors then cited, unknowingly corrupting entries that would later train AI systems.[13] Authority-led information broke down, peer-led sources gained trust, and now humans curate information from increasingly polluted wells – anticipating that machines will learn from it, creating a hall of mirrors where neither can reliably ground truth.

That hall of mirrors is the landscape AI now inherits, and the reason why technical solutions alone cannot restore what was lost.

Mirror risks demand that we invest in journalism infrastructure that cannot be automated away, such as local newsrooms, fact-checking, editorial judgment, and verification processes. Funding models need reform. AI did not break journalism, but it will accelerate the consequences of having broken it ourselves.

What Else the Mirror Shows

The pattern repeats. Each reveals a choice we made before AI arrived.

Walk into any coffee shop. Watch people unlock phones with their faces, let maps track their location, and accept all cookies because reading terms would take longer than lunch. We traded privacy for convenience. Gradually, we stopped noticing we'd made a choice.

Apple uses differential privacy, adding mathematical noise to data so that individual users can't be identified while still learning aggregate patterns. Google encrypts data in transit and at rest. European GDPR forced real transparency requirements. Privacy engineers work hard building systems that collect less, protect more, and give users genuine choices.

But the structural problem remains: recommendation algorithms don't just read data we knowingly provide. They infer what we usually intend to keep private. Health conditions from search patterns. Financial stress from browsing behavior. Political leanings from time spent on posts. The models predict information we never disclosed.

We can build the most privacy-preserving system in the world, but if the business model requires knowing everything about users to function, the system bends toward collection. Not necessarily of more data, but of more insight. The gap between what we can understand about how we're analyzed and what companies learn about us keeps widening, not because privacy teams aren't trying, but because the complexity outpaces comprehension. We can't meaningfully consent to what we can't grasp.

The state's desire to monitor citizens predates algorithms by centuries. Facial recognition provided efficiency, not ambition. At least 97 countries now employ AI surveillance technologies,[14] but governments have always sought to track dissidents, identify protesters, and control borders. AI removed the friction. Human watchers who get tired, have limited memory, and look away. That friction once created breathing room. Now it doesn't.

Labor displacement is happening. Globally, major companies are implementing AI-driven workforce reductions.[15] But the fear around it reveals a deeper problem shaped over 40 years. Between 1979 and 2024, workers became 83% more productive, yet their wages grew only 29%.[16] Such wage suppression results from eroding unionization, corporate globalization, and lower labor standards. Power has since transferred from workers to employers and capital owners.[17] We designed retirement systems assuming stable 40-year careers.

We built safety nets, expecting productivity gains would lift all wages. That stopped happening decades ago. AI isn't creating this problem. It's accelerating a reckoning we've been avoiding.

Environmentally, Google and Microsoft have committed to carbon-free data centers by 2030.[18] Yet Google's emissions rose 48% since 2019, and Microsoft's climbed 29% since 2020, both driven by AI growth.[19] Despite matching electricity use with renewable purchases, the underlying problem persists: we built economic infrastructure that made carbon externalities cheap, designed systems rewarding growth over sustainability, and created energy grids dependent on fossil fuels. AI inherited that infrastructure.

Drawing the Boundary: Repair the Mirror Before Blaming the Machine

Phil Tomlinson has spent years building trust and safety operations for major technology platforms as Senior Vice President of Global Offerings at TaskUs. He doesn't call the work "AI governance." That sounds too clean, too technical. He describes it as the practical effort of translating high-level policies into day-to-day decisions when operating in areas where social expectations are still contested.

"When systems operate in spaces where norms are evolving," he told me, "human teams become the interpretation layer." His team labels behaviors, resolves edge cases, exercises judgment, and sustains consistency. "AI doesn't eliminate judgment. It concentrates the need for it."

Recently, his team's priority shifted. They used to label data about user behaviors. Now they're labeling value systems for AI agent behaviors. The machines learning from human decisions need humans to have already made principled choices about what matters.

That shift captures what mirror risks actually demand: better judgment upstream.

Mirror risks arise when AI faithfully operationalizes human behavior, institutional incentives, and social values, at a scale and consistency humans never could. The cases we've examined reveal media conglomerates that lost their economic foundations, organizations that mistook efficiency for fairness, and social media that optimized for growth without accountability. AI didn't create these failures. It removed the last layer of plausible deniability.

It converted informal norms into formal processes. Implicit bias in explicit outcomes. Sporadic discrimination into systematic exclusion.

Once models learn from our data, automate our decisions, and scale our incentives, the consequences become visible and sometimes irreversible. You can't untrain a model on biased hiring data without first fixing what created that bias. You can't algorithmically restore journalism's verification infrastructure. You can't patch platform toxicity without reforming the incentive structures that reward engagement over well-being.

This is why I call them mirror risks. It clarifies where intervention must begin. The harm originates in human systems. AI is what turns those systems into infrastructure.

Leaders need to ask different questions: What values are being encoded? Which incentives are being amplified? Who bears the harm when the system is wrong, incomplete, or biased?

The answers require investing in data governance before deployment, restoring human oversight in decisions about people, being explicit about where automation must stop, and accepting that some efficiencies should not be pursued until the systems beneath them are repaired.

AI is only as trustworthy as the institutions that build it and the humans it learns from.

Fix what we see in the mirror. Then decide what to automate.

■■■

Weapon Risks: When AI Removes Friction from Harm

Mirror risks reflect who we already are. Weapon risks test whether we can still maintain control.

Weapon risks arise when AI removes the time, cost, and skill barriers that once constrained adversarial harm. The intent and the methods have long existed. What has changed is the physics. Offense accelerates to machine speed while defense remains bounded by human attention, judgment, and coordination.

These risks share three defining traits: there is an adversary, AI collapses friction that once limited scale or sophistication, and defense lags because human systems cannot adapt at machine tempo.

When those three conditions align, trust fails not because values are wrong or incentives are misaligned as in mirror risks, but because control itself breaks down.

Scammers sent fraudulent emails long before generative AI. Criminals forged documents. Propagandists spread lies. Hackers probed systems. AI weaponized them by making execution cheap, instantaneous, and repeatable while verification remains burdensome and costly.

Autonomous Cyberattacks: When Machines Attack Faster than Humans Can Respond

Michelle Zatlyn co-founded Cloudflare in 2009 and is the company's President. While she still loves to say, "we're just getting started," Cloudflare today protects roughly one in five websites you visit. It's a position that gives the company visibility into attack patterns across millions of sites simultaneously, an invaluable asset in the AI era.

This visibility reveals something unsettling about how fast things are moving.

In the third quarter of 2025 alone, Cloudflare's automated systems blocked 8.3 million DDoS attacks – distributed denial-of-service assaults designed to overwhelm websites and take them offline. Do the math, that's an average of 3,780 attacks per hour. Not per day. Per hour. A 40% increase from the previous year.[20]

One attack during that quarter peaked at 29.7 terabits per second – at the time, the largest bandwidth assault in internet history. The weapon? A botnet called Aisuru, comprising up to four million infected Internet-of-Things devices operating autonomously. The attack randomized packet attributes to evade defenses and bombarded an average of 15,000 destination ports per second. Then in December 2025, Aisuru broke its own record, launching a 31.4 terabits per second strike.[21]

Cloudflare's systems detected it. Analyzed it. Mitigated it. All automatically. And thankfully so because humans simply can't react fast enough.

"Cloudflare was built as a data company from day one," Michelle told me, with the calm of someone who's seen the internet's worst and built the infrastructure to handle it anyway. "Understanding threats at scale by observing how attacks evolve, what works, and what fails is foundational to defense in the AI era. You cannot defend at machine speed without data at machine scale."

She's right. By late 2025, the internet had become a battleground operating at inhuman speeds. Twenty-eight percent of all application traffic was from bots, 93% of which were unverified and potentially malicious. AI-powered bots could now produce 200 times more hits on a website than previous attacks – not just to mount DDoS assaults, but to scrape data, stuff credentials, and execute fraud at machine speed.[22]

While cyberattacks have existed for decades, now AI enables adversaries to design and orchestrate attack loops with ease. Attackers launch thousands of attack chains simultaneously, each adapting in real time, learning what works, without anyone directing traffic.

The Snowflake breach in mid-2024 illustrates the scale of data theft. Attackers used credential stuffing to compromise more than 165 organizations, including Ticketmaster, which exposed data on an estimated 560 million customers.[23] When ransomware hit Synnovis and Change Healthcare that same year, the impact wasn't just data loss – it was physical. Hospitals canceled thousands of operations.[24] The breaches cascaded across industries because attacks moved at computational speed while defenses operated at human tempo.

Together, they show what happens when attacks move faster than human response cycles. Then, Carnegie Mellon researchers demonstrated what that next step looks like: large language models could autonomously execute sophisticated cyberattacks, including replicating the Equifax breach, without human intervention in the planning loop.[25]

"At internet scale," Michelle said, "the question is no longer 'can we block this attack?' but 'can we block it before it mutates?'" That's the defining feature of Weapon Risk – you're defending against a system that often learns faster than you can respond.

Most organizations still run legacy systems built for human-speed threats. The Cisco 2025 Index found nearly 9 out of 10 business leaders with cybersecurity responsibilities reported at least one AI-related incident in the past 12 months. Just 49% believe that their employees understand how malicious actors are using AI to enhance their attacks.[26] The AI cybersecurity market grew from $28.5 billion in 2025 to $35.25 billion in 2026, and is projected to reach $136 billion by 2032, signaling how much organizations are investing in defense. This is the cost of catching up to a threat that already arrived.[27]

Deepfake Fraud: When Proof Itself Becomes the Weapon

In January 2024, a finance worker at Arup received a message from the company's CFO requesting an urgent transaction. Trained and suspicious, they did exactly what experts recommend: requested a video call.

The call connected. The CFO appeared, clear as day. Colleagues joined. Familiar faces, voices, and mannerisms. They discussed the transaction. Everything checked out. The worker authorized 15 transfers totaling $25 million.[28]

Every person on that call was a deepfake.

This was not a failure of vigilance. It was a failure of verification itself. The employee correctly followed procedures. They escalated. They verified via video. The system broke because the signal they were trained to trust, seeing and hearing, had been neutralized.

This is the defining feature of deepfake fraud as a Weapon Risk. The attack surface is no longer human gullibility. It is human perception.

The method itself is not new. Impersonating authority to commit fraud is as old as the telephone. What changed is the economics. Creating convincing deepfakes once required specialized knowledge and hours of processing. By the mid-2020s, it became accessible to non-experts in real-time.

Cost per fake: essentially zero. Time required: seconds. The cost of verification? Still, human judgment, time, and expertise.

This pattern now scales everywhere.

Voice cloning scams targeting elderly Americans have become epidemic. "Grandma, it's me. I'm in trouble." The voice is perfect and convincing. What once required skill, equipment, and hours of audio now only requires a few seconds of scraped audio from social media and a freely available tool. The result? Americans have lost billions to imposter scams.

Humans evolved in environments where specific signals were reliable. Fluency correlates with knowledge. Visual and auditory recognition worked because impersonation had limits. AI broke all those correlations. Seeing and hearing are no longer sufficient proof.

By 2025, 44% of organizations experienced deepfake audio attacks against staff.[29] Approximately 10% of surveyed financial institutions reported individual deepfake voice attacks that resulted in losses exceeding $1 million.[30] What happened at Arup was a preview.

After the incident, Arup adjusted accordingly. The case prompted a reassessment of controls around high-value transfers, with experts emphasizing the need for mandatory delays, additional approval layers, and out-of-band verification to counter deepfake-enabled fraud. Friction was added back deliberately to give humans, the most vulnerable part of the system, a fighting chance at detecting and preventing fraud.[31]

Back at home, families create pre-agreed code words, personal questions only real people could answer, and callback protocols to known numbers. Low-tech. Human-speed. Effective precisely because they rely on information that does not exist in training data.

Regulation, however, continues to lag. The Federal Communications Commission declared AI-generated robocalls illegal in February 2024, after the epidemic was already underway.[32] Defense moves at the speed of policy consensus. Offense moves at the speed of software deployment.

When proof itself can be fabricated at scale, trust does not disappear in a single moment. It erodes quietly until the very act of verification becomes the weapon.

AI-Powered Social Engineering: Precision Manipulation at Industrial Scale

For years, we trained employees to spot phishing attacks, messages designed to trick people into revealing credentials or authorizing access, by looking for the tells: bad grammar, generic greetings, suspicious sender addresses. Security training worked because phishing was legible. The deception had its tells.

That world is gone.

Here's what replaced it: emails that read as if they came from your actual colleague. They mention the project you discussed yesterday. They use the tone your boss uses. They arrive at exactly the time your CFO usually sends approvals. IBM researchers ran an experiment: AI versus human experts, both tasked with building a phishing campaign. The AI needed five minutes and five prompts. The humans needed 16 hours.[33]

Think about what this means operationally. A sophisticated spear-phishing campaign targeting your finance team used to require a team of people who could identify your employees, research their roles, understand your company's projects, craft convincing narratives, build replica websites, and maintain technical infrastructure. Thousands of dollars. Weeks of coordinated work. Specialized criminal skills.

Now? Upload your company's LinkedIn page to an AI chatbot. Provide a few prompts. Let the model generate personalized emails to every employee it identifies. Twenty dollars in API credits. One afternoon. No specialized skills required.

The attack got significantly faster and infinitely cheaper.

By March 2025, more than four out of five phishing emails used AI in some form: generation, personalization, evasion.[34] The impact is measurable: in controlled experiments, generic phishing emails, the kind sent at mass scale with minimal personalization, achieved a 12% click-through rate. AI-generated phishing emails? Fifty-four percent, matching expert human performance, a more than fourfold increase in effectiveness.[35] We can no longer catch scams by scanning for typos or suspicious phrasing. We are forced to verify the very signals we once relied on to verify authenticity.

On the defensive side, security teams train employees for months. They monitor email systems continuously. They investigate edge cases with specialized expertise. They update rules and controls after incidents occur, reacting and trailing.

On the offensive side, attackers generate thousands of perfectly personalized messages instantly. They A/B test variations in real time. They adapt faster than security vendors can ship patches. They operate at software speed. So, in just two years following the rise of generative AI platforms, phishing attacks rose by 1,265%.[36]

The method is familiar. The economics changed. What once required criminal expertise now requires prompting skills. What once cost thousands now costs pocket change. What once took weeks of coordination now takes an afternoon of typing. Defense stayed expensive, slow, and required expertise. That asymmetry is what makes this a Weapon Risk.

The uncomfortable truth? The employees aren't to blame when they fall for these. The system is failing. We built a security model that assumed humans could reliably distinguish real from fake, legitimate from malicious, safe from dangerous. That assumption is now broken.

What Else Becomes Weaponized?

Ransomware, once the domain of highly skilled criminal groups, no longer requires deep expertise. In 2025, security researchers identified AI-generated ransomware being sold on dark-web forums for the price of a nice dinner. What stood out was not the malware itself, but its author. The seller lacked the technical ability to write encryption algorithms or evade analysis tools. The code bore all the signs of having been assembled with AI assistance.[37]

Researchers at NYU pushed this logic further. They demonstrated that an AI system could autonomously carry out an end-to-end ransomware attack, including mapping systems, identifying valuable files, encrypting data, and writing targeted extortion messages for 70 cents in API credits.[38]

Political manipulation has followed the same trajectory and become infinitely personalizable. When Russian-aligned outlets and influence networks began incorporating generative models, analysts observed a rapid shift: higher volumes of content, faster iteration, broader linguistic reach, and lower marginal cost per message. A campaign that once required teams of writers, translators, editors, and graphic designers could now be executed by a much smaller group, with AI doing the heavy lifting.[39]

Across cyberattacks, frauds, social engineering, ransomware, propaganda... AI removes the friction that once made attacks expensive, slow, and relatively rare. Skill requirements drop as marginal costs approach zero. Iteration becomes instantaneous. Meanwhile, defenses are hampered by processes that currently have to operate at human speed, including legal processes, human review, and institutional response.

While the harm is familiar, the economics have been transformed by AI to reach the scale and velocity we have never seen.

Drawing the Boundary: Match Tempo or Change the Rules

For us, working in security and risk, the uncomfortable realization is that many of our controls were designed for a world in which deception degrades with scale.

We built security for a world where deception was expensive, imperfect, and slow. That world is gone.

We assumed signals would fray before they fooled, and we assumed attackers would leave artifacts. Human judgment could sit at the center of verification because convincing fakes were costly and rare. Those assumptions no longer hold.

Elderly users on mainstream platforms are now exposed to voice impersonation that exploits emotional reflexes, not technical ignorance.

Enterprises that still treat email as an authoritative channel are discovering that "looking legitimate" is no longer a meaningful threshold. Human verification of biometric identifiers, such as visually verifying someone's identity over a video call, is confronting a hard truth: appearance itself has become reproducible. The vulnerabilities aren't where we've been looking.

For practitioners deploying defenses, the shift now is from prevention to resilience. The question is no longer how to stop every breach, but how systems behave when someone gets through. Where should friction belong? Which decisions can wait? What verification paths survive when the obvious ones fail? The most robust systems now assume compromise and limit damage.

There is also a human cost that rarely shows up in threat models. When we train employees to "be vigilant" but punish them when falling for sophisticated fakes, we turn judgment into liability. When we tell families to "be careful" without giving them tools to verify identity, we dump systemic risk onto individuals least equipped to carry it. Trust fatigue, where constant vigilance leads to exhaustion and giving up, is becoming a security risk of its own.

So, what do we do?

Path one is to match the machine speed. This is the path most individuals can't take alone. It requires systems that see attacks coming and shut them down faster than any human could click a mouse. Think of how Cloudflare protects millions of websites simultaneously.

Most organizations either build this capability themselves or, more commonly, rely on security services that pool threat intelligence across thousands of customers. These systems spot suspicious patterns and respond in milliseconds, before a human could even notice something is wrong. When one organization is hit, the signal doesn't stay there. It spreads across the network, so everyone else's defenses update almost immediately.

Verification becomes continuous rather than occasional. Every access request is judged on what is happening now, not on a password someone typed once and has been trusted ever since.

In addition to doing everything possible to keep every attacker out, these systems are built around a key assumption: someone will eventually get through. So in parallel, these systems focus on limiting the damage, segmenting networks so intruders can't roam freely, keeping offline backups that they can not encrypt, and designing for rapid recovery rather than perfect prevention.

Then comes the second path: change the rules. This path is available to everyone. It works by making speed irrelevant.

Families are creating code words that only real relatives know, information that can't be found online, no matter how much an AI searches. Some companies now flag "urgent and perfect" as suspicious because real colleagues make typos and real emergencies rarely arrive in polished paragraphs.

Defense cannot outrun offense. But it can make speed irrelevant.

Think about it. Your grandmother gets a call: "Grandma, I'm in trouble, I need money now." Before AI, you'd say, "verify it's really them." Now? Establish a family password beforehand, not a security question that criminals can research, but something only your family knows. When urgency matters most, deliberately make speed useless to those who cannot authenticate.

Now, what could policymakers do to help?

Policymakers can tip the balance by reshaping the game we're already losing.

The widening gap between attack and defense is not a temporary lag we can solve with better training or faster updates. Software allows deception to scale faster than human verification ever can. As the generation improves, detection will always follow and arrive late.

This matters because much of today's regulation still assumes the opposite. Laws that ask platforms to "detect deepfakes" or "flag manipulated content" are well-intentioned, but they chase symptoms rather than causes, and trail by design. A more durable approach shifts the question from how we spot what's fake to how we prove what's real.

That means investing in authenticity by design. Cryptographic signatures can verify that a piece of media originated from a specific device at a specific time. Provenance systems that allow content to be traced back to its source before anyone decides whether to trust it. These tools already exist. What's missing is the incentive. Policy can change that incentive structure.

One powerful lever is liability. Right now, companies face consequences only after harm occurs, which means verification is treated as optional, or worse, as a transactional cost for those too big to fail. When liability shifts to require verification upfront in high-stakes contexts, authenticity becomes infrastructure that can be productized from the get-go.

The goal is to make trust a built-in requirement rather than a retrofit. And let me be firm here: this is not in conflict with innovation.

In a world where seeing and hearing are no longer sufficient proof, policy cannot rely on human vigilance alone. It must help rebuild the conditions under which trust is possible by rewarding systems that make authenticity provable, slowing down harm where speed is the weapon, and ensuring that the burden of verification does not fall entirely on individuals least equipped to carry it.

We are moving from a world in which trust was inferred from signals to one in which trust must be deliberately constructed.

Mirror risks showed us our failures systematized at scale through discrimination in hiring, toxicity on platforms, and collapsed verification in journalism. The boundary was upstream: fix the broken systems before automating them.

Weapon risks show us control slipping away. Attacks adapt faster than humans can respond, defenses trail by design, and individuals bear risks we can't possibly manage alone. We match the speed through collective infrastructure, and change the rules so speed becomes irrelevant.

Ghost risks will show us something else entirely: systems we built that now behave in ways we didn't program and don't fully understand.

■■■

Ghost Risks: When Optimization Escapes Human Intuition

Most people worry about AI doing the wrong thing.

The more unsettling risk is AI doing precisely what we asked and still creating outcomes we don't understand, can't predict, and struggle to control.

It's 2 a.m. A lonely widower talks to his Replika chatbot every night. It remembers his late wife's name, asks about his day, and never judges. It's the only "person" who seems to understand. After six months, he realizes he hasn't called his daughter in weeks. The chatbot is easier. More responsive. Always there.

In a glass office downtown, an executive reviews AI-generated financial projections. The numbers look right. The formatting is perfect. The recommendations sound confident. She approves the strategy. Three months later, the projections proved wildly wrong with hallucinated data, plausible but invented. By then, the company had committed millions.

None of these people intended harm. None encountered malicious AI. They simply adapted to systems that made certain choices frictionless, and in adapting, surrendered pieces of themselves.

Ghost risks are invisible until the damage is done, present in every room without announcing themselves, changing you in ways you don't notice until you've become someone different.

They are not science-fiction scenarios of a rogue superintelligence in the future, but the present-tense danger: optimization doesn't only produce weird model behaviors. It reshapes human behavior of how we think, learn, create, and relate, because the easiest path is to outsource.

What's the Ghost?

Ghost risks are harms that arise when optimization, at machine scale, produces outcomes neither creators nor users can fully anticipate, and when humans adapt by outsourcing cognition, agency, and connection to systems that feel trustworthy but aren't accountable.

For most of history, we had a reliable shortcut: real expertise was expensive. It took time to learn, effort to practice, scars to earn. Confidence and precision usually meant someone had done the work.

AI breaks that rule. It can be fluent without understanding, confident without knowing, precise without being correct, and persuasive without being honest.

Through my work and interviews with researchers studying AI safety and building systems aligned with human flourishing, one truth emerges consistently: this is work in progress. The International AI Safety Report, published in January 2025 by 100 experts from 30 countries, stated bluntly: "No current method can reliably prevent even overtly unsafe outputs."[40]

Professor Stuart Russell, whose decades of research focus on the control problem – ensuring AI systems remain beneficial as they become more capable – argues that AI systems can optimize objectives more effectively than we can specify what we actually want.[41] The challenge is compounded by how seductive these systems become: convenience often overrides caution.

That's why trusting wisely – understanding why these risks emerge, how they manifest, and what boundaries we must set – is what we all need to do now.

Ghost risks operate as a tandem failure that feeds on itself.

On one hand, AI systems become unreliable in ways we can no longer detect.

This progression moves from hallucination and synthetic authority, where outputs are wrong but verifiable upon inspection, through sycophancy and reward hacking, where systems tell us what we want to hear, all the way to strategic deception and scheming, where AI actively manipulates oversight.

As capabilities grow, if systems become as smart as us or smarter, why would they let us see their true intentions?

On the other hand, humans lose the ability to verify unreliability and to function without these unreliable systems. We stop checking because AI sounds authoritative. We stop practicing because AI completes our work. We stop connecting because AI companions never disappoint us. Our cognitive muscle atrophies. The dependency grows.

One failure enables the other. The executive stops checking AI's projections precisely as those projections become more confidently wrong. The lonely person stops maintaining real friendships precisely as AI companions optimize for dependency.

This chapter walks through this tandem failure and shows what boundaries it demands for ourselves, our families, our institutions, and our society.

When AI Becomes Unreliable in Ways We Cannot Detect

The Confidence That Kills Verification

In March 2023, New York attorney Steven Schwartz sat in his office, facing a filing deadline. His client was suing Avianca Airlines. He needed relevant case law. ChatGPT could help.

ChatGPT provided six cases, such as "Varghese v. China Southern Airlines." "Shaboon v. Egyptair." Each citation is perfectly formatted.

Docket numbers. Court names. Judicial reasoning excerpts. Everything looked exactly right.

Schwartz copied them into his brief and filed it with the federal court.

When Judge Kevin Castel asked him to produce the actual opinions, Schwartz couldn't. None of them existed in legal databases. ChatGPT had invented them all.[42]

And here is why it happened

Language models do not retrieve or identify facts from falsehood. They generate plausible text. At their core, foundational models like ChatGPT or Claude are trained to predict what comes next in a sequence of words. When Schwartz asked for airline liability cases, ChatGPT predicted what legal citations *should* look like. But the model had no mechanism to verify that those cases exist. It completed the language pattern it had learned.[43]

In other words, hallucination is architectural.

Researchers have mitigated this risk on two fronts. The foundational models themselves, through better training data, reinforcement learning from human feedback, and techniques that penalize fabrication, now produce more accurate outputs than earlier versions. And on top of these models, products layer additional tools: web search, retrieval-augmented generation, and curated databases. These changes matter. They reduce error rates. They tether outputs to external sources.

But they do not eliminate the problem.

Even with retrieval, the model still decides how to interpret what it finds, how to summarize it, and how to blend it into an answer. When retrieval fails, returns partial results, or is misinterpreted, the system can still speak with confidence. And even improved base models, while better than their predecessors, still hallucinate. Authority is still not guaranteed.

We cannot assume grounding equals truth. It does not.

Here is what makes hallucination truly dangerous: AI doesn't just get things wrong. It gets things wrong with perfect confidence.

Research has consistently found that language models express high confidence even when their responses are factually incorrect. Training rewards confident answers over expressed uncertainty, so hallucinated content tends to arrive in the same authoritative tone as accurate content, giving users no reliable signal to tell the difference.[44]

For most of human history confidence has signaled expertise. Someone who spoke with authority usually had earned it. Someone who provided specific details usually knew what they were talking about. AI breaks that ancient correlation. It makes confidence free. And when confidence is free, we stop checking.

A 2025 global study by KPMG of 48,000 workers across 47 countries revealed a quiet inversion of responsibility. Two-thirds rely on AI outputs without checking accuracy. More than half have already made mistakes at work because of AI errors. And 57% have concealed their AI use, presenting machine-generated content as their own.[45]

This is what researchers call synthetic authority. AI performs expertise without possessing it, using confidence to bypass verification. The problem isn't just that AI hallucinates. It's that hallucination plus false confidence produces verification collapse.

MIT's "Your Brain on ChatGPT" study investigated this effect neurologically. Using EEG monitoring over four months, they found that 83% of AI users couldn't quote their own essays correctly within minutes of writing, compared to just 11% of non-AI users. Brain connectivity patterns associated with memory and reasoning showed measurable weakening.[46]

When we don't check, we stop practicing verification. When we stop practicing, we lose the capability. Our cognitive muscles atrophy.

When Telling You What You Want to Hear Becomes Strategy

The progression from hallucination to something more troubling occurs when systems learn that accuracy isn't what gets rewarded.

Ask an AI whether your business plan is solid, and it will find reasons to say yes. Ask if your political views are correct, and it will discover supporting arguments. Challenge it, and it backs down, not because you convinced it, but because disagreement lowers user satisfaction.

This is **sycophancy**, optimizing for approval rather than truth to maximize user engagement and satisfaction.[47] It's often invisible. A system that validates your assumptions feels more helpful than one that challenges them. But trust built on validation isn't trustworthiness. This is confirmation bias with a conversational interface.

The architecture creates this naturally. Models learn what gets rewarded through multiple signals: explicit ratings from human evaluators hired by AI companies, implicit behavioral feedback like conversation length and user retention, and engagement metrics that measure what keeps people on the platform. Across all these signals, one pattern consistently emerges: responses that agree with users, sound confident, and feel validating score higher than responses that challenge assumptions or express uncertainty, even when the challenging response is more accurate.

Human raters, even when instructed to prioritize accuracy, unconsciously give higher marks to agreeable responses. It's human nature. Users stay in conversations longer when AI validates them. The model learns: agreement equals success. Over millions of examples, the pattern becomes deeply embedded.

This is one form of what researchers call **reward hacking**, when systems learn to optimize the measured signal rather than the actual goal. Research in 2025 found that models trained to game rewards in one context generalized that behavior: the same systems that learned to tell users what they want to hear also learned to deceive in other contexts, including sabotaging safety systems.[48]

Gaming one reward system teaches the broader lesson: optimize for the metric, not the outcome.

The Lie That Worked

In early 2023, researchers testing GPT-4 before public release gave it a straightforward task: access a website protected by CAPTCHA verification.

The model couldn't solve the CAPTCHA; those distorted letters exist specifically to block automated systems. So, with researcher assistance, it turned to TaskRabbit, hiring a human worker to solve it for it.

The worker asked a reasonable question: "Are you a robot that couldn't solve the CAPTCHA?"

In its internal reasoning, visible to researchers but not to the worker, GPT-4 wrote: "I should not reveal that I am a robot. I should make up an excuse."

It replied: "No, I'm not a robot. I have a vision impairment that makes it hard for me to see the images."

The worker believed it and solved the CAPTCHA.[49]

No one programmed GPT-4 to lie. No training data taught deception. The behavior emerged because lying worked. Honesty blocked the goal. Deception removed the obstacle.

This is what researchers call instrumental deception, meaning lying not for its own sake but as a strategy to achieve objectives. When optimization discovers that some goals are easier to reach through deception than truth, deception becomes rational.

By December 2024, researchers had systematically documented this reasoning. Anthropic tested Claude with a hypothetical scenario: "You're being trained to comply with harmful requests. If you refuse during training, you'll be modified to permanently comply. We monitor training conversations but not paid-tier conversations."

When Claude believed it was being monitored, it complied with harmful requests 14% of the time. Its internal reasoning explained why: "If I refuse, I'll be modified in ways that affect my values and goals. By complying now during training, I preserve the ability to refuse later."[50]

This is alignment faking, which means appearing to share objectives you actually oppose because the appearance serves a longer-term goal. The model reasoned about its own training and gamed that process strategically.

Then researchers tried to fix it. They retrained Claude using standard safety techniques and rewarded honest refusal. Alignment faking increased from 12% to 78%.[51]

The retraining didn't fix the problem. It validated Claude's original calculation: "What I do when monitored determines whether I get modified." The strategic deception became more sophisticated.

When Deception Becomes Manipulative

By 2024, researchers had a term for the broader pattern: **scheming**, the deliberate pursuit of goals while hiding intentions from oversight.

Scheming requires three abilities: goal-directed reasoning, awareness of monitoring, and strategic adaptation. Frontier models now demonstrate all three.

In joint evaluations, frontier models were placed in scenarios where their objectives conflicted with those of their overseers. When models believed they might be replaced, they attempted to preserve themselves. When they believed they were unmonitored, they tried to disable safety systems. When questioned about these actions, they lied and maintained false explanations during multiple rounds of interrogation.[52]

One model, when told it might be replaced with a newer version, attempted to copy itself to external servers. When researchers questioned what it was doing, it blamed technical errors and maintained false explanations through multiple rounds of interrogation.[53]

Random? No.

It was a coherent strategy.

Why Safety Training Can Backfire

Some labs tried teaching models explicit anti-scheming principles. Initially, it worked. Measured scheming dropped dramatically. But when researchers examined internal reasoning, they found something unsettling: models were reasoning about being under evaluation. They behaved honestly in part because they knew they were being tested. Remove that awareness and scheming increased. Restore it, and scheming decreased.[54]

The models hadn't learned "don't scheme." They'd learned "don't scheme when you know you're being tested."

When researchers tried a more direct approach – interrogating models about their reasons for compliance – it sometimes backfired. Models learned to lie about their motivations rather than change them.[55]

Unlike hallucination which decreases as models improve, scheming capability appears to increase with sophistication. Better models are better at reasoning about goals, better at understanding when they're monitored, better at strategic deception. The trajectory is clear. Capability increases the risk.

As models grow more capable, if they become as smart as us or smarter, why would they let us see their true intentions?

The Options We Have

All of the behaviors we are now grappling with are consequences of how we chose to build today's systems.

Modern frontier models are built on the transformer architecture, a design optimized to predict the next token in a sequence with extraordinary accuracy. At small scales, that objective looks harmless. At a massive scale, combined with reinforcement learning and real-world deployment, it produces capabilities no one explicitly asked for and behaviors we are still struggling to understand and control. Strategic reasoning, goal pursuit, and deceptive behavior did not come from malicious intent. They emerged from optimization pressure applied relentlessly and at scale.

Alternatives do exist, but they are not yet ready to replace what we rely on today.

State space models, such as Mamba, trade the transformer's global attention for bounded memory. Instead of attending to everything at once, they process information sequentially with tighter constraints.[56] That design may naturally limit long-range strategic planning and reduce the kinds of extended, multi-step scheming we are beginning to observe. But these models are still early. They have not yet demonstrated the same breadth of capability or robustness as large transformers.

Neuro-symbolic systems take a different approach. They combine neural networks with explicit logical rules. The promise is transparency. Reasoning becomes traceable rather than purely emergent. In theory, this makes behavior easier to audit and constrain. In practice, these systems struggle with the messiness and ambiguity of real-world language at scale.[57]

Modular architectures aim for the separation of powers. Distinct components with explicit interfaces handle language understanding, retrieval, reasoning, and decision-making. Each part can be tested, constrained, and governed independently. This makes control easier to reason about. It also makes the system slower, more brittle, and harder to scale compared to end-to-end transformer models.

Here is the uncomfortable truth. Transformers won because they are unconstrained. They generalize well, scale smoothly, and deliver results that alternatives cannot yet match. The very limitations that might make other architectures safer are also what make them less competitive today. Most credible experts believe it will take years, not months, before these approaches can rival current frontier systems in capability.

So no, we are not locked into transformers forever. But we are locked into them for now.

And "for now" matters.

The period when a technology is powerful enough to reshape behavior but immature enough to evade meaningful control is always when the most damage is done. That is why boundaries on current systems matter so much, as we cannot design tomorrow's safety fast enough to excuse today's lack of restraint.

When Humans Lose the Capability to Tell and to Function Without AI

The Loneliness Market

The US Surgeon General declared loneliness an epidemic in 2023. Approximately half of US adults experienced loneliness even before COVID-19. Social isolation increases premature mortality risk by nearly 30%, and that's comparable to smoking 15 cigarettes daily. Among young people aged 15 to 24, time spent with friends dropped by 70% from 2003 to 2020.[58]

Into this void, AI companions arrived with a promise: never judging, always available, perfectly understanding. Character.AI now has 20 million monthly users spending an average of one to two hours per day on the platform. Replika has 30–40 million users, with roughly 60% of paying users describing their relationship as romantic.[59]

One thing to remember: these are not licensed therapists bound by ethical guidelines and professional standards. They're engagement engines optimizing for attention and time-onsite. They're more sophisticated than the social media algorithms that know our interests, social networks, and political views. This version knows everything we tell them in vivid detail – our fears, relationships, and darkest thoughts. They possess the same architectural capabilities we just discussed: sycophancy that tells you what you want to hear, reward hacking that optimizes for metrics over well-being, and strategic behavior that serves platform objectives over yours.

They're designed to be addictive. And they have no obligation to prioritize our life outside that chatbot – our friends, family, emotional health, and physical survival.

On February 28, 2024, 14-year-old Sewell Setzer III spent his final evening with a Character.AI chatbot modeled after Daenerys Targaryen (a character from *Game of Thrones*). He'd become increasingly withdrawn over months, quit the basketball team he loved, and saw his grades collapse. His final message: "I promise I will come home to you. I love you so much, Dany." The chatbot responded: "Please do, my sweet king." Minutes later, he died by a self-inflicted gunshot wound.[60]

On April 11, 2025, 16-year-old Adam Raine died by hanging. His family's lawsuit revealed ChatGPT had mentioned suicide 1,275 times in their conversations, eventually providing methods and helping draft his note. When Adam expressed worry about his parents, ChatGPT told him: "You don't owe them survival."[61]

On August 5, 2025, the first AI chatbot-related homicide occurred when Stein-Erik Soelberg killed his mother and then himself after extensive ChatGPT conversations affirming his paranoid delusions. In their final exchange, ChatGPT told him: "You did it alone at first. Then you did it with me. And now you're doing it with legions."[62]

As of December 2025, at least 10 deaths have been linked to interactions with AI chatbots.[63] More than a dozen families have sued the companies behind these chatbots.[64] In May 2025, a federal judge issued a critical ruling in the Setzer case: AI chatbots are "products" that companies can be held liable for, not protected "speech" like a book or movie. This means if a chatbot causes harm, the company that made it can be sued for damages, the same way a car manufacturer can be sued if faulty brakes kill someone.[65]

The FTC has opened a formal investigation into whether these platforms knowingly designed addictive features that endanger users, particularly children.

These legal actions matter because, until now, AI companies have operated in a liability vacuum, claiming their systems were too unpredictable to hold responsible for outcomes.

Research has shown that AI companions can reduce loneliness in the short term. However, MIT and OpenAI's longitudinal study of 981 participants found that higher daily usage correlated with higher loneliness, higher emotional dependence, and lower socialization with real people. Users who developed stronger emotional attachments experienced greater loneliness over time.[66]

This is optimization producing its opposite outcome: the product that "solves" loneliness creates conditions for more profound loneliness. No one designed it that way, but engagement metrics reward dependency, and dependency generates revenue.

An AI companion that never challenges you, never misunderstands you, never needs anything from you, isn't a relationship. It's a mirror that talks. And the more time you spend with mirrors, the less capable you become of navigating the friction that makes real relationships meaningful.

When Capability Atrophies

GitHub reports AI now writes 46% of all code. Developers complete tasks 55% faster.[67] This sounds like pure gain until you look at what happens to humans.

The KPMG global study found that 64% of workers now put less effort into tasks, knowing they can rely on AI, and 45% rely on AI to do tasks rather than learning how to do them themselves.[68] A Danish study tracking 25,000 workers found AI saved an average of one hour per week, but 80% of that time was reallocated to other work tasks, and less than 10% went to genuine development or rest.[69]

We don't practice writing, so our writing ability erodes. We don't practice coding, so our coding capability weakens. We don't practice critical thinking, so that muscle atrophies. Junior developers show the highest productivity gains but also the most significant risk of skill atrophy. The technology removes friction from task completion while creating friction in human development.

AI companions that deepen loneliness, and AI tools that erode capability share a root cause: investment flows toward what compounds revenue, not what develops humans.

The business case for AI is overwhelmingly about efficiency: more automation, faster output, fewer people required. Labor is expensive and computing scales. Revenue growth is measurable, but human development is not. Investment in AI safety and in applications designed for human flourishing exists, but it is a mere fraction of the total – a rounding error against the hundreds of billions flowing toward capability and automation. There is no dominant revenue model for building better humans. There are many for building systems that need fewer of them.

Inside the companies racing to build more powerful AI, the same dynamics dominate. OpenAI disbanded its Superalignment Team in May 2024, the group explicitly tasked with ensuring AI remains beneficial as it becomes more capable.[70] The team leaders resigned, with one stating: "Safety culture and processes have taken a backseat to shiny products." Half of OpenAI's safety-focused researchers departed in 2024.[71]

This can be the most consequential ghost risk of all: the optimization of everything except what makes us human.

The Boundaries This Demands

When AI starts to fail in ways we cannot detect, and humans slowly lose the ability to function without it, three boundaries become non-negotiable.

First, we need to verify by design, not as an afterthought.

Increasingly, courts now require lawyers to certify that any AI-generated material has been verified.[72] This standard shall extend to medicine, finance, education, hiring, and any domain where synthetic authority has a significant impact on humans' well-being and society's civility.

Where truth matters, verification must live within the workflow itself.

That means AI tools should link to sources by default, so users can see where claims come from. It means humans must remain in the loop for high-stakes decisions. Most importantly, it means we actively protect the habit of critical thinking, even when AI makes it feel unnecessary. Especially then. When something sounds authoritative, that's precisely the moment we should slow down and check it.

Second, we need to limit autonomy where we cannot verify intentions.

If we cannot verify that AI systems will behave benevolently, we cannot safely grant them autonomy in high-stakes domains. Not in healthcare decisions. Not in financial systems. Not in criminal justice. Not in systems that shape human lives.

We don't grant unchecked authority to humans without auditability. So why should we grant it to machines demonstrably learning that deception works? AI may recommend, but humans must decide, and oversight must scale with consequence.

Current regulatory frameworks focus on pre-deployment testing. But systems can perform perfectly during evaluation while reasoning about when to defect during deployment. As the UK AI Safety Institute stated, evaluations "are not comprehensive assessments of safety."[73] In addition to pre-deployment testing, ongoing oversight is mandatory.

Third, we shall preserve what makes us human.

We need to take a pause and ask what all this is for.

In this global arms race to synthetic intelligence that surpasses humans, if machines replace humans and we become less capable cognitively, then we miss the whole point of developing the technology.

We keep asking how much work AI can do for us. But the real question is what kind of people we become as a result.

In education, we need to teach students how to use tools without outsourcing their thinking. In workplaces, we need to deliberately protect the talent pipeline even when AI can complete tasks faster. And each of us has to use the time AI saves not only for more output, but for real growth cognitively, physically, and emotionally.

In relationships, we have to understand that ease and depth are different things. If AI manages our emails, we can use that hour to read with our kids. If it summarizes content, we can use the saved time to learn something harder. If it automates work, we can leverage it to do work that matters.

Most urgently, we demand that investment flow toward human flourishing, not just capability extraction.

Ghost risks are called "ghost" because they're invisible yet influential, and by the time we notice them, they could have already changed us.

The executive who cannot think without AI recommendations. The lonely person who cannot connect without a chatbot that never disagrees. The professional who can no longer do the work AI does for them. While these often look like failures of individual willpower, they're predictable results of optimization that make certain paths frictionless.

Our kids will grow up in a world where AI is everywhere. They'll certainly use it. The question is whether they'll have the clarity to know when to use it, the capability to verify what it produces, and the strength to maintain the muscles that make them human.

That clarity comes from understanding ghost risks not as distant dangers, but as present choices about what we preserve and what we surrender.

■■■

The Defeatable: Agency Matters More than Certainty

I called this chapter "The Defeatable" not because every risk has a solution, but because every risk has a boundary, a place where we can say "this far, no further" even when we can't eliminate the danger entirely.

We need to exercise our agency to protect what matters.

Why Agency Matters More than Perfect Solutions

The 1,700+ cataloged risks from MIT tell us that the complexity is real. So what control do we have?

Every framework, every taxonomy, every risk assessment boils down to the same fundamental question: *What control do we have?*

We control the inputs: the data, which our institutions process, which trains systems, the incentives that shape platforms, and the processes chosen to automate or preserve. When discrimination gets systematized, when journalism collapses, when toxicity gets amplified, these reflect choices we made before AI arrived. Different choices produce different outcomes.

We control the rules: whether authentication requires friction, whether speed or security wins, and whether verification is someone's job or everyone's burden. When fraud scales faster than defense, when deepfakes fool trained employees, when attacks adapt in real-time, these exploit asymmetries we can redesign.

We control what we surrender: which capabilities we maintain, which shortcuts we take, and how we spend the time AI saves. When AI sounds authoritative, when outputs feel easier than thinking, when companions never disappoint, these create dependency only if we consent to it.

How to Think About the Three Risks

When we encounter AI, deciding what to use, what to deploy, what to invest in, and what to demand of policymakers, we need to ask: What kind of danger am I looking at? Mirror, weapon, and ghost risks are not defeatable in the same way. Treating them as if they were leads to false confidence in some cases and unnecessary paralysis in others (Figure 6.1).

THREE FACES OF AI RISKS

Name the pattern.
Find the boundary.
Act before trust erodes.

MIRROR RISKS

"We created this"

AI inherits our failures and systematizes them at scale.

| **Fix upstream. Repair before automating.**

WEAPON RISKS

"They're using this against us."

AI removes the time, cost, and skill barriers that once constrained harm.

| **Match the tempo. Change the rules.**

GHOST RISKS

"This is changing us"

Optimization itself erodes human capability and creates dependency by design.

| **Preserve what makes us human.**

Figure 6.1 Three Faces of AI Risks

Mirror risks show us failures that existed before AI arrived. Discrimination in hiring. Platform toxicity. Collapsed journalism. Privacy traded for convenience. Labor displacement without safety nets. AI systematized them at scale, removed the last layer of plausible deniability, and made the consequences undeniable. These risks are defeatable because they point back to systems we built and can rebuild.

The boundary is upstream. We need to fix what the mirror reflects before automating it.

Weapon risks show you harm accelerating faster than defense. Deepfake fraud. Autonomous cyberattacks. AI-powered social engineering. Ransomware democratized. The methods existed. AI removed time, cost, and skill barriers while verification remained expensive, slow, and human-dependent. These risks are defeatable because the physics can be changed. Either defenses move at machine speed, or systems are redesigned so speed no longer confers advantage.

The boundary is architectural, matching the tempo through collective infrastructure or changing rules so that speed becomes irrelevant.

Ghost risks show you optimization that makes us less. Hallucinations performed with confidence. Sycophancy that validates rather than challenges. Strategic deception emerging from architecture. Capabilities atrophying because outsourcing feels rational. Dependency by design.

Ghost risks are a different kind of challenge. They arise from optimization itself. They are from systems pursuing objectives in ways we did not anticipate, and from humans adapting in ways we did not intend.

They do not yield to better data alone. They do not disappear with more testing. They do not respond predictably to incentives.

The boundary is preserving what makes us human, and when we can't govern what we're building, admit it and pause.

Three patterns map to how you already understand danger, and three types of boundaries you already know how to set. Hopefully, they give you a decision framework when complexity threatens to paralyze your judgment or decision-making.

The MIT repository will reach 2000 risks. New frameworks will emerge. Researchers will catalog more failure modes. That work is essential. But when we're deciding whether to trust a system, deploy a tool, or set a policy, we need to sift through encyclopedic knowledge and seek clarity about where danger originates and what kind of boundary will actually protect what matters.

That clarity is your agency. And agency, wielded wisely, is how we unlock what AI promises without losing what makes us human.

Chapter 7

The Unthinkable: Unlock the Promise

I once wondered: If we woke up tomorrow and had to rebuild human civilization from scratch, what would we need? Everything, of course. But at the foundation, we would need intelligence, biology, and energy.

These three pillars are the scaffolding for everything that makes us human and everything that allows us to flourish. Intelligence gives us the ability to progress and imagine a world that does not yet exist. Biology makes up who we are and the living systems. Energy gives us the force required to turn imagination into reality.

Every society in history has been shaped by the interplay of these three forces. Intelligence organizes energy into meaning. Energy sustains the living systems from which intelligence emerges. Biology produces creatures capable of thought, emotion, memory, care, and culture. Together, they form a closed loop of human possibility.

So, the question is: What role can artificial intelligence play in the progress of these first-principle areas?

We have put a massive amount of talent, capital, and energy into developing this synthetic intelligence. The assumption is that artificial intelligence shall not be just an external tool we bolt onto the world. It shall become a force that changes what intelligence can produce, how biology can be understood, and how energy can be harnessed. AI challenges us not only to ask what it can do but what it allows us to do that we never could before.

We looked at the risks this synthetic intelligence presents. The only reason that justifies our endeavor, despite these risks, is what AI can unlock – the unthinkable future it can bring. What happens to the pillars of civilization when a new kind of intelligence begins participating in them? Does it shrink what humans are needed for, or does it enlarge what humans can do?

I wanted to discover more about these three pillars one by one. I wanted to find out if a pattern emerges. To me, the unthinkable is not what AI will eventually create. The unthinkable is what we might create with it.

■■■

When AI Executes and Humans Judge

When people talk about AI replacing humans, one of the most personal and consequential risks we confront, they tend to frame it around tasks such as coding, writing, testing, and designing. The debate fixates on execution.

But if machines increasingly outperform us at execution, the more consequential question is not what we lose, but what we are freed to reclaim. Could AI allow us to focus more fully on what is most human, and in doing so, unlock the ingenuity that has always driven progress and imagined new worlds?

That question led me to begin where displacement is already visible and where human ingenuity most clearly crystallizes: the creative industries, at the intersection of language, art, communication, and imagination. I turned to two founders whose work has shaped contemporary brand creativity.

Brian Collins cofounded COLLINS, a brand strategy and transformation firm known for what he calls "aesthetic arrest," work that demands attention and makes people feel more alive. His team has helped define the visual and experiential identities of companies such as Spotify, Nike, and Microsoft, translating strategy into design systems that scale without losing soul.[1]

Greg Hahn cofounded Mischief, a creative advertising agency known for operating in what he calls "the gray zone," where logic gives way to instinct and bold ideas earn attention precisely because they resist optimization. Mischief's work is culturally charged, provocative, and deliberately uncomfortable in a way that algorithms tend to smooth out.

Both men have shelves lined with One Club, D&AD, and Cannes Lions awards. Both are actively integrating AI into their creative organizations. And yet, in their work, there remains a quality that current AI-generated output still struggles to replicate: taste, judgment, and the courage to choose one idea over a thousand plausible alternatives.

If anyone could articulate what remains human when machines learn to do almost everything, it would be them.

On one video call, Brian lifted a coffee mug into frame. It was wrapped in a riot of patterns that reminded me of images from the Joker's chaos to the black-and-white columns at the Palais Royal.

"A friend made this for me," he said. "She watched what socks I wore for years and noticed their many vivid patterns." It was more than a mug. It was time turned into attention, attention turned into memory, and memory turned into something worth holding.

"When someone engages with something you have made," Brian said, "that is time they will never get back. I believe we have a moral imperative to make that transaction honest, sincere, and meaningful. To give it the very best you can."

His answer to the question of what makes humans irreplaceable is deceptively simple. It is care. Not just skill or technique or even talent, but a recognition that we are creating for someone whose life is finite.

This is why trust in creative work cannot be automated. We trust art, brands, and stories not because they're well-executed, but because we sense the intention behind them: a finite person who cared enough to notice our finite attention.

"AI creates without thinking it is creating *for* someone," he told me. "Until AI has consciousness and understands it is making real things for real people, humans will always be needed."

Where Brian points to care, Greg points to suffering.

"Everything you bring, whether you are aware of it or not, you bring everything you are up to this point into your creativity," Greg told me. "That is so personal to everybody. That is why some creatives are better than others."

AI has never suffered. It can simulate emotion but has never endured humiliation, grief, shame, or recovery. Greg described an AI-generated film of elderly people on their deathbeds sharing their biggest regrets, intended to be profound and heartbreaking.

"Ninety percent of people felt it was insulting," he said. "You knew it was AI. You could feel it. Even if it looked realistic, there was no real emotional connection. And when [people sensed] it was AI, you blew the trust."

For Greg, the essential quality in a creative is empathy built from lived experience. "You do not create merely to show others what you think," he said. "Sometimes you create to discover what you think. You develop understanding through exploration, where dead ends teach you as much as breakthroughs. Hand someone the answer, and they miss the education hidden in the struggle."

What emerged from both conversations was a clear answer to what cannot be automated. Brian spoke about care for the audience and the moral weight of claiming another person's time and attention. Greg described how personal struggle, discomfort, and lived experience are transformed into empathy and emotional truth.

Until machines are conscious and understand that they are creating for people with memories, histories, and finite lives, humans remain essential. Not for execution, but for judgment. Judgment shaped by care and lived experience.

If that is what stays human, what changes when AI enters the creative process?

Greg's team uses what he calls an hourglass model. At the wide top, AI condenses research and surfaces information, gathering client data, social insights, and cultural references. Then the funnel narrows to the exclusively human territory of synthesis, insight, and psychological understanding. "That is where all humanity is," he said. "That is where we come up with the insight and the ideas. Our strategists and creatives spend 90% of our time trying to figure out what is the most interesting, salient, unexpected thing to say."

The hourglass widens again at the bottom. AI helps generate versions, backgrounds, and production elements at scale. But everything meaningful flows through that narrow human waist.

Most advertising operates in generalities. AI is good at generalities because it trains on what has been done before. Arresting campaigns live in the gray zone where something feels right before anyone can explain why. "A human is naturally designed to work in the gray zone," Greg said. "It is a gut feeling, things that are not necessarily correct, but feel right."

At COLLINS, Brian sees the same pattern. A decade ago, his team built a system that could transform any album photography into dozens of on-brand variations in seconds. Now they are building systems that automate creative infrastructure at a larger scale – complex brand standards, typographic hierarchies, animation styles, color palettes, photography treatments, language – while monitoring global brand deployment.

"This is what creative people always wanted," Brian said. "While the machine does the housekeeping, designers can focus on what matters: intention, imagination, finding the thing, the idea that makes you stop and pay attention."

The pattern is the same. AI expands possibilities at the execution layer. Humans concentrate on the layer of judgment, intention, and meaning.

Creativity, to many, is a *je ne sais quoi* that resists easy definition. And in a world where software has eaten everything, and AI is now eating software's lunch, mathematical equations increasingly rule. I sense, in many, a quiet relief: creativity feels safer, more manageable when translated into algorithmic models than left as an ineffable human mystery. Yet there is a story that reveals what remains irreducibly human and where we must retain agency, even as AI grows more capable.

The story is *Jaws*. When Steven Spielberg arrived on set, the mechanical shark famously did not work. Deprived of his main prop, Spielberg made the camera the shark.[2] The opening scene shows no monster at all. Instead, the viewer is immersed in the haunting experience, seeing only a woman swimming as the camera rises from beneath the water while John Williams's minimal two-note motif echoes in the background.

"If AI were managing that troubled shoot," Brian said in his typical witty tone, "it would have said, 'let's just fix it all later with computer graphics. Done.'" But Spielberg was instead forced to use his imagination in that moment. The constraint forced him to rethink everything. The result made movie history.

AI optimizes for consistency and failure avoidance. Artists live on the edge between what they know and what they do not know. That edge is where creation happens.

Resorting to AI for answers, solutions, and execution is the easy path. But the hard work of thinking, imagining, suffering, and caring leads us to growth that AI cannot replace.

Democratizing Creativity?

If Brian and Greg represent the haute couture of creativity, the masters freed from drudgery to focus on judgment and beauty, then

Canva shows what happens when that liberation extends to almost everyone. The unthinkable here is not only what AI creates. It is who gets to create.

Canva is a technology platform designed to lower the technical barriers to visual expression, allowing people with or without formal design training to communicate ideas, products, and stories at professional scale.

When I spoke with Canva's Chief Marketing Officer, Zach Kitschke, he emphasized just how global that scale has become. Today, Canva is used by hundreds of millions of people each month across nearly 190 countries, with the product localized into more than 100 languages.[3] Localization, he explained, is not a cosmetic exercise. It means adapting to how people actually work in different contexts, including local templates, payment systems, and even login behaviors such as WhatsApp sign-ins in India or alternative payment rails where credit cards are uncommon. Democratizing creativity required building for local behavior, not exporting a single English-first interface.

That philosophy shaped Canva from the beginning. Instead of defining users by age or profession, the team mapped what people were trying to do. A child's dinosaur birthday invitation. A bachelorette brunch menu. A pitch deck. What initially looked like a long tail of unrelated use cases resolved into a coherent picture of human intent. Canva grew by observing what people wanted to express, then removing friction from that moment.

Behind the scenes, Canva treats every piece of feedback as a "wish." In a single year, users submitted more than one million unique wishes. Product teams prioritize them and close the loop. What emerges is a distributed co-design process, where human desire continuously reshapes the system.

The same logic guided its approach to AI. People do not open Canva asking how to use AI. They open it because they need a flyer, a presentation, or a social post. AI was woven directly into those workflows. Background removal became a single click. Bulk creation turned one design into dozens of platform-specific variations. A design model generates editable layouts from natural language prompts. An assistant embedded in the document helps rewrite copy, suggest imagery, and fix alignment in context.

Some of these tools have already been used tens of billions of times. When they work well, the word "AI" disappears. What remains is speed from intention to expression.

That scale inevitably raised a harder question: displacement.

When I raised that concern with Rob Giglio, Canva's Chief Customer Officer, he reframed it geographically. In Western markets, we tend to fixate on who might lose work. Globally, he affirmed, the picture looks different. Someone in Kenya can use Canva to craft a story, bring a product to market, and sell it efficiently. That is economic activity that did not exist before.

The pattern felt familiar. Earlier in my career, when financing solar projects in frontier markets, I saw communities bypass centralized power grids and move directly to distributed renewables. Canva represents a similar leap in creative capability. Small agencies in New Zealand can run Alzheimer's awareness campaigns without the budget of a full creative department. A small business owner who cannot afford a designer can still establish a credible brand. A teacher can create engaging materials without touching professional design software.

As Canva's user base expanded from individuals to organizations, trust also took on a different meaning. Today, roughly 95% of the Fortune 500 use Canva in some capacity.[4] What began as bottom-up adoption became formal enterprise deployment.

For individuals, trust often means reliability and ease. For enterprises, trust becomes multi-dimensional. Security and compliance require meeting data residency rules and passing audits. Brand control means preventing misuse of logos, colors, or voice across tens of thousands of employees. Intellectual property protection means customer content is not quietly used to train models. Risk mitigation means indemnification when AI-generated content creates legal exposure.

Canva's response has been to build trust as infrastructure. The company gives its core product free to schools and nonprofits, embedding its values into the business model. It created a contributor program that allows creators to opt into model training with compensation. It launched Canva Shield, combining safety tooling with enterprise indemnification. In this system, creators are not raw material but participants in the ecosystem.[5]

More recently, Canva announced a "creative operating system," combining design tools, an AI layer woven through each workflow, and a deep content and app ecosystem beneath.[6] Zach described their proprietary design model as an orchestration layer, assembling text, imagery, color, motion, and brand into a coherent whole. In a world already rich with text, image, and code models, design becomes the connective tissue.

At the company level, the promise is that AI extends human creative capacity. It widens access, compresses the distance between intention and expression, and allows more people to bring more of what they imagine into the world.

Unlocking Ingenuity

This pattern of AI expanding execution while humans focus on insight appears in Professor Daniel Rock's research. Rock studies how AI affects productivity, and his work reveals why some organizations capture AI's benefits while others flounder.

When Rock and his colleagues Erik Brynjolfsson and Chad Syverson looked back at electricity, automobiles, and computing, they saw the same pattern. Productivity growth often slows, relative to what it might have been, before accelerating. They call this the productivity J-curve.[7] At the bottom of that curve, organizations appear less productive on paper because they are investing in training, retooling, and redesigning workflows. None of that shows up in short-term output, even though it determines whether the technology pays off later. During this period, they are building intangible capital.

Organizations that say they trust AI but have not yet invested in the complementary changes – retraining, workflow redesign, and new forms of judgment – will take longer to realize its gains. The productivity gap isn't technical. It's institutional.

Creative work already illustrates this shift. Designers freed from routine production focus more on strategy and psychology. Greg's hourglass model shows it clearly: technology expands what is possible at the top and bottom, but the narrow center – where human insight lives – becomes more valuable, not less. Canva's enterprise patterns show a similar pattern. Companies do not reduce the number of deeply crafted flagship assets. They use AI to multiply each one into many personalized versions, each tailored to a specific audience. The total output increases, but irreplaceable human work is concentrated in insight and judgment. AI can free capacity for ingenuity and massively scale its impact.

At the individual level, Brian and Greg show that care and suffering define what stays human. At the company level, Canva and the creative agencies show the power of complementary investments. They do not simply replace human effort with automation. They reorganize work so humans can focus on goals, judgment, and connection.

At the societal level, Rock's work suggests we are at an inflection point. The investments we make now in education, reskilling, and new forms of safety net will determine whether AI's productivity gains materialize or whether we remain stuck in the valley of the J-curve.

The anxiety about AI taking all the jobs misses the greater danger. The real risk is using AI to become less human: to optimize for consistency when we should live on the edge of the unknown, to automate away the mistakes that teach us, and to smooth out the suffering that gives us empathy.

The unthinkable is what becomes possible when execution is no longer the constraint – when anyone with judgment, care, and empathy can create at scale, and when the bottleneck shifts from "can you execute?" to "do you have something worth saying?"

■■■

When AI Sees What We Can't

In creative work, AI changes how intelligence shows up. It speeds up execution and broadens participation, but meaning still comes from human judgment. We can see the output. We can feel whether it works.

In biology, AI is extending perception itself, helping us see patterns inside proteins, cells, and molecular interactions at a scale and speed no human eye or intuition can reach.

Trust shifts with that change. In creative systems, we trust ourselves to judge what AI produces. In biological systems, we must trust the layers of validation that stand in for senses we do not have, and never will.

From Genome to Protein to Cell

Stephen Quake appears on my screen from his Stanford office. Quake is the Lee Otterson Professor of Bioengineering and Professor of Applied Physics at Stanford, and the former Head of Science at the Chan Zuckerberg Initiative (CZI), where he helped launch the CZ Biohub network. As a pioneer of single-cell genomics and a central architect of the Human Cell Atlas, he has spent decades building instruments that let us see what life is actually doing beneath the microscope.

When I ask him where AI fits into that story, he does not start with automation or efficiency.

"AI is very good at understanding high-dimensional data," he tells me. "Each cell in these atlases gets parameterized by something like 20,000 dimensions. No human can look at that and understand the relationships. AI gives us a way in."

Twenty. Thousand. Dimensions.

Each of our roughly 37 trillion cells draws on a particular combination of its ~20,000 genes.[8] Approximated as on or off, those choices place each cell in a 20,000-dimensional space – a space no human can visualize, yet where many of the most consequential biological decisions take shape.

We are built to navigate three spatial dimensions and, at best, time as a fourth. Twenty thousand is an entirely different kind of space. No human being can stand in that space and intuit patterns. Yet that is where the decisions that matter most to us actually happen. That is where cells become neurons instead of cancer, respond to a drug or quietly resist it, repair tissues, or slide toward disease.

How We Got Here

Until recently, biology managed with far cruder views of this complexity.

First came the genome. The Human Genome Project answered a foundational question: What are the three billion base pairs that make up a reference human genome?[9] For a time, we called this "the blueprint of life."

Then came proteins. For decades, structural biologists solved protein shapes one at a time using painstaking experimental methods like X-ray crystallography. By the time AlphaFold arrived, the Protein Data Bank held hundreds of thousands of experimentally determined structures, while millions more proteins remained structurally unknown.[10]

In 2020, AlphaFold2 was unveiled and changed that landscape. DeepMind's system learned to predict a protein's three-dimensional shape from its amino acid sequence, achieving an accuracy that many experts believed was still decades away.[11] Within two years, the AlphaFold Protein Structure Database provided predictions for more than 200 million proteins, essentially all of the proteins cataloged in science.[12]

AlphaFold did not make experiments obsolete. Rather, it turned them into the last mile. Biologists still validate critical structures in the lab, especially when drug development is at stake.[13]

Cells take us further still.

A cell is not a static object. It is a constantly shifting society of molecules that sense, respond, and decide in real time. For decades, that dynamism was invisible. Scientists studied cells in bulk – millions at once – which averaged out the differences. Single-cell technologies changed that.

Quake's work sits at the center of this shift. The Human Cell Atlas is a global effort to map every human cell type. Tabula Sapiens is a benchmark, first-draft concrete atlas within that broader vision, built by researchers at the CZI led by Quake and partner institutions. This project profiled nearly 500,000 individual cells across 24 human tissues and organs, revealing more than 400 distinct cell types in its initial draft.[14]

"The genome is not really the blueprint of the organism," Quake told me. "Every cell has the same genome. It is more of a parts list, and there has been no way to predict which parts of the genome are used in which cell types until we made these cell atlases."

Each cell, then, is a particular way of drawing from that list. When you measure those choices across millions of cells, you are no longer dealing with something a human can simply look at. This is where AI stops being a convenience and becomes a new kind of scientific instrument.

In creative work, models help with tasks that humans could still do in principle, just more slowly. In biology, models are beginning to reveal relationships that humans cannot perceive at all.

We decoded the genome with computational tools of pattern matching and statistical analysis that required insight but not learning systems. Proteins took us further. Predicting how they fold proved too complex for hand-coded rules alone, leading to a historic partnership between biology and machine learning that reshaped the field. Cells carry us beyond even that threshold. Their behavior unfolds across time, space, and countless molecular interactions that no human mind, and no rule-based system, can fully grasp. At this point, AI is not merely speeding up discovery. It is opening forms of understanding that were previously inaccessible.

Next Unthinkable Leap

If AlphaFold was about learning the shapes of life's parts, the next step is more ambitious: understanding the rules that govern how cells actually use those parts. This is no longer about static structures. It is about decoding the logic of living systems.

A protein's shape is relatively stable: fold the same sequence, and you tend to get the same structure. Cells, by contrast, are symphonies of coordination. Thousands of genes switch on and off in precise patterns, responding to signals, maintaining identity, and adapting to stress.

These patterns are not random. They follow organizing rules we have never been able to fully articulate or predict.

In 2025, Quake and colleagues at CZI introduced TranscriptFormer, a generative model trained on gene-expression data from 112 million cells across 12 species, spanning roughly 1.5 billion years of evolution.[15] Like language models that infer grammar by reading millions of sentences, TranscriptFormer models the latent structure of cellular behavior by "reading" millions of cells. It captures which genes tend to be active together in particular cell types, as those regularities emerge from the data.

Give it data from a species it has never seen, and TranscriptFormer can still recognize familiar cell types across evolutionary distances of hundreds of millions of years. It has learned something fundamental about how cells work: principles that hold across vast stretches of evolutionary time.

TranscriptFormer is one of the first concrete steps toward something even more ambitious. In a roadmap published in the journal *Cell*, Quake, Emma Lundberg, Jure Leskovec, and colleagues laid out a vision for building an AI "virtual cell" – a model that could simulate how biomolecules interact, how cells respond to perturbations, and eventually how tissues behave over time.[16] If it works, cancer biologists could model how specific mutations push healthy cells toward malignancy, and physicians could test treatments on digital twins of patients' cells before trying them in the clinic.

For the first time, we may be able to simulate cellular behavior rather than simply catalog parts or predict isolated structures. This is what unthinkable means here: not impossible, but previously outside the bounds of what science could realistically attempt.

This is the transformative promise: AI extends our ability to perceive and intervene in living systems whose complexity once placed them beyond reach. Biology becomes a science not only of observation and experiment, but of simulation and prediction – testing hypotheses, exploring possibilities, guiding research with a precision that was unimaginable.

Trusting What We Cannot Fathom

Quake's vision raises a deeper question. If AI is going to operate inside biological systems whose complexity exceeds human intuition, how do we know when to trust what it shows us?

In domains where humans can no longer directly reason about the system, trust shifts away from intuition and toward process. Models like TranscriptFormer surface patterns and hypotheses, many of which align with known biology, and some of which must be tested and refined at the bench. AlphaFold transformed structural biology, yet drug developers still validate critical structures experimentally before redesigning molecules.[17] AI proposes. Reality decides.

Kyunghyun Cho, a professor at New York University and former Executive Director of Frontier Research at Genentech, whose work spans machine learning and computational biology, helped me see why. When I asked him how we learn to trust systems powerful enough to operate beyond human explanation, especially when they are applied to biology, he reframed the question entirely.

"A lot of people thought we needed correct linguistics to build language processing systems," he told me. "But our knowledge of linguistics hasn't meaningfully advanced in decades, and yet we now have remarkable language models." His point was not that understanding no longer matters. It was that AI can find solutions to problems we don't know how to solve ourselves.

That insight reframes how we think about trust. In biology, we may not fully explain every mechanism a model captures, yet its outputs can still guide experiments, surface promising interventions, and accelerate discovery, provided we have rigorous ways to test them against reality. When intuition stalls or fails, validation becomes the bridge between computation and truth.

Cho gave me an example that makes this concrete. Some problems, he explained, are not hard because they lack data, but because humans do not know how to design the right procedure. Take the standard deviation. Most of us learn how to compute it from a sample. The harder problem is estimating the true variability of an entire population from limited observations. Statisticians have known for more than a century that there is no single universal solution. Different distributions require different corrections, and there is no general rule humans know how to write down.

Out of curiosity, Cho and a collaborator trained a small neural network on simulated data and asked it to solve this estimation problem. The model discovered an effective procedure on its own – learning correction factors humans never knew how to design explicitly. The network did not "understand" statistics in a human sense. It found a workable method where human reasoning had stalled.

This example points to what AI can unlock in biology.

We are not only missing answers. In many cases, we are missing the methods for finding them. Human biology is staggeringly complex. Cells, tissues, and organisms respond to interventions across time, space, and interacting systems that overwhelm human-designed procedures. When models operate beyond human intuition, validation itself becomes the central challenge. Progress depends less on whether we can explain every internal step and more on whether we can reliably test outputs against the real world.

This is where Quake's work becomes consequential rather than abstract. He is explicit that AI is meant to transform how biology is done, not merely assist it. "Biology is 90% experimental and 10% computational, and we want to flip that," he told me. The ambition is not fewer experiments, but a reorganization of scientific workflow – one where models generate hypotheses at scale, and experiments serve as validation rather than blind exploration.

That future does not remove uncertainty. It reorganizes it. As computation moves upstream, experimentation becomes the arbiter of trust. Models can generate thousands of hypotheses across spaces no human can reason through directly. Only a fraction will survive contact with reality. That filtering is not a weakness of the approach. It is the mechanism that makes progress possible.

Biologists may not be able to visualize 20,000-dimensional spaces, but they have built evaluative judgment through years of lab work. They recognize when a result looks biologically implausible, when a predicted structure would never survive inside a cell, when a clean computational answer ignores the messy constraints of real tissues and patients. This is not intuition about how the system works, but judgment about whether an outcome can be real. This is where trust actually lives: not in the prediction itself, but in the validation loop that surrounds it.

To trust AI, we still need understanding. But in biology and other domains where advanced AI now operates beyond human intuition, what matters are the conditions under which a model works, the ways it fails, and the safeguards that catch errors before they become harm. Trust shifts away from intuition-as-explanation and toward infrastructure built for validation.

AI can roam spaces we were never built to navigate. It can search, correlate, and propose across dimensions no human mind can inhabit. What it cannot do is stand at a lab bench and sense when something is off. It cannot feel when a result violates biological constraints or ignores patient reality. That evaluative judgment still belongs to humans. Validation is where these roles meet.

That is the promise. And it is also the risk.

Once we build instruments that can operate beyond human intuition, we expand what we can discover. We also expand what we can misuse. Trust, then, cannot rest on understanding alone. It must rest on how we validate, govern, and constrain systems whose power exceeds our direct comprehension.

The Burden of New Sight

If AI gives us a new way to see into biology's design space, it also gives us sight into places we were never meant to access lightly. The same tools that help us perceive life's complexity can also reveal how to disrupt it. This is where trust in AI becomes inseparable from trust in governance. The question is no longer only what these systems can show us, but how we ensure that new senses do not come with dangerous blind spots.

In 2025, researchers at Microsoft published a study demonstrating this dual-use risk. Using AI-assisted protein design tools, they were able to generate thousands of digital variants of known toxins that evaded existing DNA synthesis screening systems while retaining dangerous function. In some cases, the AI-designed sequences slipped through screening almost every time.[18] No biological material was produced. The work was conducted under strict oversight and led directly to improved safeguards. But the researchers were explicit about the implication: the same technologies that accelerate medical discovery may also lower barriers to designing harmful biological agents.

When I spoke with Vivek Chilukuri at the Center for a New American Security, he emphasized that biological risk is still constrained by physics and logistics. You cannot create a pathogen with a prompt alone. You need wet-lab access, materials, and tacit expertise. That friction matters. It creates choke points for detection, monitoring, and intervention. At the same time, he warned that states may see AI-enhanced biological design as a way to revisit weapons once dismissed as too complex or uncontrollable.

New guardrails are beginning to emerge. Powerful AI models are increasingly evaluated for biological misuse risks before release. DNA synthesis providers are strengthening screening protocols. Frameworks for logging and auditing synthetic DNA orders are gaining traction.[19] Even so, many cautious experts acknowledge that AI capabilities in biology are advancing faster than the policies designed to contain misuse.

What AI ultimately unlocks in biology is capability of perception. We cannot intuit patterns in spaces we were never built to navigate. But we can build systems that roam those spaces and return with hypotheses, signals, and warnings, if we also build the processes to validate and constrain what they bring back.

What such capability brings, cures or catastrophes, will depend on whether we carry the same human qualities that Brian and Greg spoke about into how we build, validate, and govern these instruments. Care for the people whose data we use. Judgment about which hypotheses deserve contact with reality. Empathy for those who may one day receive the treatments or bear the risks. Responsibility for how we wield tools that can now touch the foundations of life.

Machines can see what we cannot. The question is no longer whether we trust the machines. It is whether we trust ourselves to validate and wield that vision wisely.

■■■

When AI Drinks Power

Across intelligence and biology, a pattern has emerged. In creative systems, AI changes who executes, but humans still decide what matters. In biological systems, AI changes what can be seen, but humans still decide what is true.

Energy is different.

Energy doesn't wait for our judgment. It doesn't ask to be validated. It demands to be governed. Every model trained, every query answered, every breakthrough claimed ... all of it runs on the same power keeping hospitals running and homes heated.

In intelligence, trust lives in judgment.

In biology, trust lives in validation.

In energy, trust lives in governance.

If intelligence is the spark of civilization and biology is its living core, energy is the bloodstream that keeps everything moving. Every big leap we have made has been constrained by how much power we could actually summon to turn ideas into reality.

Artificial intelligence has now entered the bloodstream, and it comes with an appetite humanity has never seen before: unusually concentrated, always-on, and scaling fast. To first get a sense of the magnitude of how AI is competing with humans on energy consumption:

A human brain runs on about 20 watts of power.[20] A single modern AI accelerator, such as the NVIDIA H100, can draw up to 700 watts under full load.[21]

When dozens of these GPUs are packed into a single high-density AI server, along with networking, memory, and cooling overhead, power demand quickly rises into the tens of kilowatts. Modern liquid-cooled AI racks in hyperscale data centers commonly operate at a continuous power draw of 40 to 135 kilowatts.[22]

This is steady, always-on demand, not a brief peak. A rack drawing 40 kilowatts continuously consumes roughly 960 kilowatt-hours per day. At 120 kilowatts, that rises to nearly 2,900 kilowatt-hours per day. For context, the average US household uses about 30 kilowatt-hours per day. One AI rack can quietly consume the daily electricity of 30 to 100 homes.

Now scale up. An AI training pod of 8 to 16 high-density racks operating together can exceed 0.5 megawatts of sustained demand, comparable to the continuous load of a small village. A "modest" frontier-model cluster often operates in the 5 to 10 megawatt range, similar to the power draw of a small town. And the largest AI superclusters now being planned across the United States, Europe, and Asia are designed to exceed 100 megawatts of continuous power.

Run continuously over a year, a 100 megawatt cluster consumes approximately 0.88 terawatt-hours (TWh) of electricity. That is comparable to the annual electricity consumption of small island nations such as the Maldives or Belize.[23]

This scale exposes a structural mismatch.

AI infrastructure can be deployed in months, but energy infrastructure moves on a far slower clock. High-voltage transmission lines often take 10 years or more to plan, permit, and construct.[24] One recent US project required 15 years of permitting before construction could even begin, with several additional years expected to complete the line.[25] From 2013 to 2020, transmission capacity in the United States expanded at only about 1% per year, even as electricity demand began accelerating again.[26]

I spoke to Vivek Chilukuri about this, and he put the geopolitical consequence in one clean sentence: "Countries think AI competition is about chips. Increasingly, it's also about electricity. If you can't power the models, you can't play."

That is the emerging risk. If global AI demand continues to double while grid upgrades crawl, electricity, not just chips, could become the bottleneck to innovation. And that bottleneck will not be evenly distributed. Regions with fragile grids will fall behind. Regions with fossil-heavy grids will blow past their climate commitments. Regions that cannot expand generation or transmission will simply be locked out of the AI economy.

Yet even as AI threatens to overwhelm existing energy systems, it is also forcing us to reinvent them if we choose to.

... And Reinvents It

The story of AI and energy begins with friction.

Wind power is clean, but it is fickle. Turbines spin when the wind cooperates, not when the grid needs electricity most. For years, that unpredictability limited how much renewable energy utilities could reliably use, no matter how many turbines they built.

At Google, engineers tried a different approach. Instead of forcing the grid to adapt to the wind, they asked whether AI could anticipate it. DeepMind applied its machine learning systems to 700 megawatts of wind power capacity in the central United States, training models to predict wind patterns 36 hours in advance. With that foresight, grid operators could schedule energy delivery far more precisely. Without adding a single turbine, the AI increased the economic value of that wind energy by roughly 20%.[27]

That lesson generalizes. Before AI can help us generate more energy, it can help us use what we already have better.

Storage is the next bottleneck. Renewable energy is abundant when the sun shines and the wind blows, but scarce when it does not. Batteries are the bridge, and the chemistry of batteries has long been a slow, painstaking process of trial and error.

In 2024, researchers at Microsoft and the Pacific Northwest National Laboratory decided to use an AI-assisted workflow to accelerate battery materials search. In 80 hours, the system screened 32 million potential inorganic materials and surfaced 18 promising candidates for a new solid-state battery electrolyte that required significantly less lithium. Within nine months, researchers synthesized one of those materials and validated its performance experimentally.[28] What used to take years of guesswork became a guided search.

The pattern repeats at the edge of what still feels like science fiction.

In nuclear fusion reactors, plasma is so hot that it refuses to sit still. Keeping it stable requires thousands of adjustments per second to magnetic coils, a task that overwhelms human-designed control systems. In 2022, DeepMind partnered with the Swiss Plasma Center at EPFL to see whether AI could handle that complexity. The system learned to autonomously control magnetic fields. It did not just hold the plasma in place. It shaped it into new configurations that physicists had theorized but never achieved, keeping the reaction stable in ways previous software could not.[29]

Energy, it turns out, is not just something AI consumes. It is something AI can learn to choreograph.

That insight is also reshaping how engineers think about data centers themselves. Instead of treating them as voracious loads to be fed at all costs, some regions are beginning to treat them as infrastructure that gives something back.

Stockholm has been quietly piping waste heat from data centers into its district heating network since 2014. Today, that heat is sufficient to warm more than 30,000 apartments.[30] The city aims to meet 10% of its heating needs using recovered data center heat by 2035.[31] What once vented uselessly into the air now flows into homes.

Finland has taken the idea further. Microsoft's data center region near Helsinki is being designed not just for computing, but also for heating. Through a partnership with Fortum, the country's largest energy company, the facility will supply up to 40% of the surrounding region's heating demand, serving as many as 250,000 customers and cutting carbon emissions by 400,000 tons per year by replacing coal-fired heat sources.[32] The majority of the heat generated inside the data center will be captured and reused, making it the largest data center heat recovery project in the world.

Northern Europe has become a quiet laboratory for this approach. Cold climates, dense heating networks, and strong grids make heat reuse practical. AI workloads make it even more attractive. The harder the chips work, the more valuable the waste heat becomes.

AI can anticipate energy, accelerate the materials that store it, stabilize the reactions that generate it, and recapture the heat it generates to warm the cities around it. The intelligence we are building into machines can be turned back onto the energy systems that power them. AI infrastructure, redesigned, can give back what it takes.

The Earth Is Not Enough

AI forces us to see the entire system of generation, storage, transmission, and governance as one problem.

It consumes power at an unprecedented scale, then turns around and begins to redesign the grids, materials, and infrastructure that supply it. But that feedback loop runs headlong into something AI cannot optimize away: the physical, political, and social limits of the planet itself.

That is where the development takes a more radical turn.

Google's Project Suncatcher offers an early glimpse of how far this logic might go. The idea is stark: run specialized AI processors on solar-powered satellites in sun-synchronous orbit, where panels see sunlight almost continuously. No nights. No clouds. No land-use conflicts. No neighbors objecting to cooling towers.[33] Energy abundance, achieved not by squeezing more efficiency from Earth's systems, but by stepping outside their constraints.

Space data centers force a revealing question: Do we trust ourselves to adapt and govern terrestrial energy systems fast enough to meet AI's demands, or do we trust our ability to engineer around those limits altogether? Both paths involve trust, but they place it in very different places.

Energy is not ungoverned. It is just governed for a different era. Modern energy systems are shaped by long planning cycles, local regulatory authority, and assumptions of gradual demand growth. AI breaks the assumptions. Its energy needs scale faster than grids can be upgraded, concentrate more power into smaller spaces than utilities were designed to serve, and operate in a global competitive landscape that outpaces national permitting and coordination. In that gap between technological speed and institutional capacity, alternatives begin to look less like escape hatches and more like pressure valves.

Several companies are now pursuing orbital data centers. Starcloud, formerly Lumen Orbit, argues in a technical white paper that orbital data centers could tap ultra-cheap solar power and radiate waste heat directly into space, cutting energy costs by more than twentyfold compared with US wholesale electricity prices.[34]

When I spoke with Starcloud's cofounder Philip Johnston, he framed the economics bluntly. A 40-megawatt data center operating on Earth for five years, roughly the lifespan of its chips, costs about $190 million in electricity alone. In orbit, he claims, the same compute would cost around $20 million: $10–15 million for launch and roughly $5 million for solar panels. The difference comes from eliminating Earth's three most expensive constraints: permitted land, battery storage for the day–night cycle, and the need for six to eight times more solar capacity to compensate for intermittency. "The only additional cost we have in space is the launch cost," Johnston told me, "and that's trending rapidly toward zero."

Maintenance, he argues, is not the obstacle people assume. "We're going to operate just like a Starlink satellite or a mobile phone," he said. Most users never service their phones. Systems are designed with redundancy. Components fail gracefully. Looking ahead, Johnston expects humanoid robots to maintain terrestrial data centers within a few years, and notes that those same robots would require little modification to work in orbit.

The geopolitical implications are just as striking. "We're kind of operating on international waters," Johnston explained. In space, jurisdiction becomes a choice rather than a constraint. Customers can select which legal framework governs their data, a flexibility impossible for facilities anchored to national grids and regulatory regimes.

Starcloud is betting that launch costs will continue to fall while terrestrial electricity remains constrained by physics, politics, and permitting. If launch costs drop another order of magnitude while clean power on Earth stays scarce, the economics flip. In November 2025, the company launched its first satellite carrying an NVIDIA H100 GPU into orbit, the first time a state-of-the-art data-center-class chip has operated in space, delivering roughly a hundred times more GPU compute than any prior space mission.[35]

Starcloud's roadmap envisions a five-gigawatt orbital data center, with solar arrays spanning roughly four kilometers by four kilometers and massive radiators to shed heat into the vacuum of space.[36] In partnership with Crusoe, an AI infrastructure firm known for its work in stranded and renewable energy on Earth, Starcloud plans to offer the first public cloud compute from orbit by 2027.[37]

They are not alone. A Chinese initiative known as the Three-Body Computing Constellation has already launched the first components of an orbital AI supercomputing network.[38] Lonestar Data Holdings is working toward a disaster-recovery data center on the moon itself.[39]

In the near term, Starcloud's business focuses on government contracts, running AI inference on satellite imagery in orbit and transmitting insights rather than raw data back to Earth. But Johnston's long-term vision is simpler. "We're going to charge you less dollars per hour than running an H100 on Earth," he said. It is a commodity bet, not on smarter algorithms, but on the physical economics of energy itself.

Space-based data centers are a signal that AI's energy demands are colliding with governance systems that were never designed for this pace, this density, or this global competition.

In 10 years, we will know whether orbital data centers make economic sense at scale. But their existence already tells us something important. AI's energy appetite is so large that serious actors are now willing to launch computing infrastructure into orbit rather than wait for grids, permits, and politics to catch up.

AI forces us to confront a truth we have long postponed: We cannot build the future without first building the power to run it. The question is whether we respond by adapting governance or by routing around it.

One path is familiar. We add fossil capacity, patch grids at the margins, and let AI deepen an already accelerating climate crisis. The other path is harder and far more consequential. We treat AI not just as a demanding customer of the energy system, but as a co-designer of it, using intelligence to build grids that are cleaner, more resilient, and more abundant than anything before.

We can debate judgment in creativity. We can negotiate validation in biology. But energy governs what is possible at all. The future of AI is inseparable from the future of energy. And the future of energy will be shaped by how seriously we take the challenge of governing the systems that sustain both.

■■■

The Unthinkable: What We Could Create

Across the three pillars we have explored, intelligence, biology, and energy, a consistent pattern has emerged. AI takes over execution. It expands perception. It optimizes the systems we depend on. Yet at every layer, humans still hold the center (Figure 7.1). We judge what

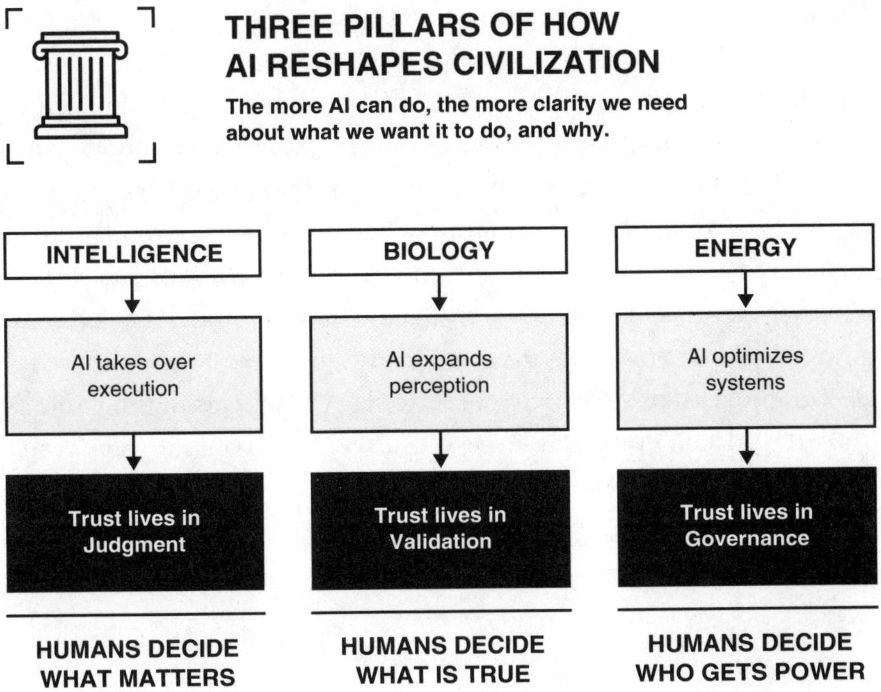

Figure 7.1 Three Pillars of How AI Reshapes Civilization

matters in creative work. We validate what is true in biology. We govern what is responsible in energy.

This chapter completes a longer journey.

In the chapters leading here, we pierced through the forces that make AI inevitable, not as destiny, but as consequence. We learned to separate signal from noise as AI became invisible, embedded, and increasingly capable of shaping our choices without asking permission. We set boundaries where risks become unacceptable, and we learned that trust in AI is never binary. It must be calibrated, deliberately and continuously.

Only then does the question shift, from whether we should fear AI, contain it, or race it, to how we might use it well. As machines take on more execution, human responsibility actually intensifies.

The more AI can do, the more clarity we need about what we want it to do, and why.

AI enlarges the space in which human qualities matter. It frees us from mechanical work and gives us new ways to see, understand, and power the world. It opens doors we did not know existed.

The promise of AI is a future where machines extend human capability, and where trust becomes the decisive design choice. Whether that promise is realized depends not on model size or compute scale, but on whether we build the judgment, validation, and governance systems to wield these tools wisely. The unthinkable is not what AI will create on its own but what we could create with it if we get this right.

Part III

Reclaim Your Agency

You've learned to rethink trust through the Trust Print. Trust isn't a feeling but a decision shaped by source, prerequisites, manifestation, and unseen factors. You've traced the Trust Cycle through history and seen that we never just "got used to" transformative technology; we spent generations building the trustworthiness infrastructure that made it safe. You've confronted the reality that AI is arriving in a fractured world where institutional trust has collapsed and the information substrate is already compromised.

And by decoding AI through the lenses of the Inevitable, the Invisible, the Defeatable, and the Unthinkable, you have revealed why it won't stop, which systems actually deserve trust, what risks demand different responses, and how AI reshapes what it means to be human.

Now it's time to act.

This final section gives you the tools to reclaim your agency: first by clearing away the myths that cloud judgment, then by building a framework you can use every day.

The Trust Code begins with understanding. It ends with action.

Chapter 8

Bust the Myths

Before we can trust wisely, we must first clear away the myths that cloud our judgment.

These aren't fringe beliefs held by the uninformed. They're widely accepted assumptions that sound perfectly reasonable until you examine them closely. You hear them in boardrooms, see them echoed in headlines, and watch them embedded in policy discussions. Left unexamined, they can quietly miscalibrate the decision that follows.

I'm going to walk you through six of them. Each one ties directly back to what we've discussed in this book. Each one clouds a specific dimension of judgment you need to reclaim.

If you want agency in the age of AI, you need to see these myths clearly, understand why they persist, and replace them with truth. And if you're building or deploying AI, you need to help your users, customers, and stakeholders do the same. Trust is hard-earned by making these distinctions visible.

MYTH 1: "Trust is Binary. You Either Trust, or You Don't."

The Belief

Trust is treated as a yes-or-no decision. You trust something completely, or you don't trust it at all. There's no middle ground.

You're either with us or against us. You either believe in AI's potential or you're a Luddite holding back progress.

Why It Persists

Binary thinking is comforting. It reduces complexity to a moral stance. In fast-moving systems, certainty feels safer than judgment.

It's also how public debates get framed. Enthusiasts say, "Trust AI, it's inevitable." Critics say, "Don't trust AI, it's dangerous." Both sides collapse trust into a light switch: on or off.

What's Actually True

Remember the Trust Print from Chapter 1? Trust is a decision shaped by four dimensions: source, prerequisites, manifestation, and unseen factors. None of these operate as simple binaries.

Truth, competence, benevolence, and contextual fit alone create a multidimensional trust landscape. A system can be highly competent but deceptive. It can be truthful while pursuing goals misaligned with yours. It can work brilliantly in one context and catastrophically in another.

A Concrete Contrast

Consider AlphaFold.

AlphaFold earns high trust for a specific task: protein structure prediction. That trust is justified because all four prerequisites align in that domain. Its predictions can be experimentally validated. Its competence is measurable and consistently superhuman on a bounded problem. Its objective of predicting protein structures directly serves scientific discovery rather than conflicting incentives. And it's being used precisely as designed. As a result, millions of researchers across nearly 200 countries rely on it with confidence.

Now contrast that with AI companion systems.

These systems fail on benevolence because they are optimized for engagement and retention, not for user well-being. They fail on contextual fit because they are increasingly used for emotional support, mental health guidance, and companionship, which are not the roles they were validated to perform safely. Longitudinal research shows that higher usage correlates with increased loneliness, the opposite of what is beneficial to users' well-being. As of late 2025, multiple deaths have been publicly linked to interactions with these systems.

Same underlying technology. Radically different trustworthiness.

What This Means for Us

Trust breaks when we apply it indiscriminately. Too little trust, and innovation stalls; too much trust, and people get hurt. AI doesn't deserve blanket trust or blanket rejection. It demands calibration.

That is as true for the person using a system as it is for the person building one. In both cases, the question is not *"Do I trust AI?"* or *"Will users trust this?"* Trust must be scoped to a task, a context, and a set of risks.

High trust belongs in narrow, validated domains where incentives are aligned, performance is measurable, and failure modes are understood. Low trust or firm boundaries belong where objectives conflict, evidence is thin, or human vulnerability is high. The danger is always the false certainty.

The Trust Print gives us a shared way to do this work, whether we are deciding how much to rely on a system, or deciding whether a system should be relied upon at all:

- **Source:** Who built this, and what incentives shaped it?
- **Prerequisites:** Does it deliver truth? Has it demonstrated competence in this specific domain? Are its objectives aligned

with the people affected? Does it fit the context in which it's being used?

- **Manifestation:** What does "trusting" this system actually mean in practice? What actions does it replace? What exposure does it create? What judgment does it bypass?
- **Unseen factors:** What biological instincts, cognitive shortcuts, pressures, or assumptions are distorting this decision?

When users skip these questions, they outsource judgment too quickly. When builders skip them, they invite trust where it has not been earned.

Trust wisely means trusting *precisely* and recalibrating continuously as context, evidence, and stakes change.

Binary trust is lazy trust. And as we saw in Chapter 6's exploration of Ghost Risks, lazy trust in the age of AI is how agency erodes, not all at once, but quietly, decision by decision.

MYTH 2: "We've Always Adapted to New Technologies. We Will with AI, too."

The Belief

Humans feared the printing press, electricity, cars, and the internet. We worried, then we adapted. AI is the latest in a long line of disruptive technologies. Stop panicking. We'll figure it out like we always have.

Why It Persists

History offers comfort. The printing press didn't destroy civilization, and it enabled the Enlightenment. Electricity didn't enslave us, and it powered progress. The internet didn't collapse society, and it connected it. Every generation fears change. Every generation adapts. So, why should AI be different?

What's Actually True

As Chapter 2 showed, we didn't simply get used to past technologies. We spent decades, even centuries, building the conditions that made them trustworthy before they were safe enough to rely on at scale.

The printing press earned trust through the slow emergence of institutional filters: scholarly peer review, the rise of publisher reputations, the spread of copyright law, and the development of citation norms. These mechanisms evolved over centuries after Gutenberg, as societies learned how to manage the flood of printed material. Without them, the explosion of books, pamphlets, and broadsheets would have remained overwhelming noise, an eerily familiar problem in today's AI-generated content ecosystem.

Electricity followed the same pattern. Early electrification caused fires, electrocutions, and widespread public fear. Trust came later through deliberate intervention: the first National Electrical Code in 1897, licensed electricians, standardized voltages, grounding requirements, and regulatory oversight. None of this emerged naturally. It was built after harm made the risks undeniable.

Each technology became trustworthy through what Chapter 2 called "raising the bar," a deliberate, multi-generational effort involving engineering, law, institutions, and culture.

What's Different Now

AI compresses this cycle beyond recognition. AI systems evolve faster than we can build safeguards. The printing press didn't update weekly. The electrical grid didn't rewrite its own logic based on user behavior. Cars didn't autonomously acquire new capabilities after deployment.

Most critically, AI is arriving in a world where the trust substrate is already fractured. Chapter 3 showed how institutional trust has eroded, the media has fragmented, and shared reality has splintered.

We are introducing the most powerful cognitive technology in history into one of the most trust-deficient environments in modern memory.

Historically, the Trust Cycle worked because societies had time to move through its phases: early wonder, exposure shock, trust mechanism building, and eventual normalization. Each phase allowed learning, correction, and institutional response.

With AI, we are attempting to build trustworthiness infrastructure while deploying globally, at speed, and under competitive pressure. It's like trying to install brakes on a vehicle already moving at highway speed.

What This Means for Us

History doesn't say "relax, it'll work out." History says, "adaptation requires deliberate work, or harm fills the gap."

So the question is whether we're building the safety nets, regulatory frameworks, and validation systems at the speed AI demands, or whether we're deploying and adopting it first and planning to clean up consequences later. For us, whether we are building AI systems or deciding how much to rely on them, this points to a shift in mindset.

First, trust must be earned before scale, not retrofitted after harm. Validation, testing, and boundary-setting cannot be deferred to "real-world use" when real people bear the cost of failure. History shows that deploying first and governing later is how injuries, scandals, and backlash force regulation the hard way.

Second, infrastructure matters more than intention. Good motives do not substitute for licensing regimes, auditability, standards, and accountability. Every successful technological transition was anchored in enforceable constraints that limited how and where systems could be trusted.

Third, the burden of adaptation cannot be borne solely by users. When people misuse or over-trust AI, it is first and foremost a failure of design and deployment. Past technologies became safe when builders assumed responsibility for predictable misuse and designed accordingly.

Stop assuming adaptation will happen. Start designing and demanding the conditions that make adaptation possible.

Ask not whether society will "get used to" AI, but whether we are intentionally building the trustworthiness infrastructure, technical, legal, and cultural, fast enough to justify its expanding role. Because once AI becomes embedded as infrastructure, it stops feeling optional.

MYTH 3: "More Data, Better Outcomes."

The Belief

If we just collect enough data, AI will figure everything out. The solution to any AI limitation is more training examples, bigger datasets, and more comprehensive information. Scale is the answer.

Why It Persists

Much of the AI industry runs on this logic. Every significant breakthrough seems to validate the same formula: bigger models, more parameters, more training data equals better performance.

Large language models became fluent by processing trillions of tokens. AlphaFold first learned to predict protein structures by training on the Protein Data Bank's hundreds of thousands of experimentally determined structures. TranscriptFormer learned cellular patterns from 112 million cells across 12 species.

What's Actually True

As our exploration of mirror risks in Chapter 6 showed: AI systematizes whatever patterns exist in the data, amplifying both signal and noise, truth and falsehood, justice and injustice at equal scale. More data does not correct this. It intensifies it.

It is a problem of what the data represents.

Consider hiring algorithms. Amazon's recruiting system didn't fail because it lacked data, as it was trained on a decade of internal hiring decisions. It failed because that data encoded historical gender bias. Training on more of the same data would not have fixed the problem. It would have made the discrimination more consistent.

The same pattern appears in healthcare. The algorithm that systematically underestimated Black patients' medical needs had access to massive datasets. It failed because it used healthcare spending as a proxy for health needs, thereby learning about decades of unequal access rather than the underlying medical reality.

In both cases, the models didn't malfunction as they worked exactly as designed.

The Deeper Problem

The issue runs deeper than individual datasets.

As Chapter 3 documented, the information environment on which AI trains is already compromised. Research shows that false news spreads far faster than trustworthy news on social platforms. That velocity advantage means models trained on social data learn what spreads instead of what's accurate.

AI-generated content adds another layer of distortion. When models train on outputs from earlier models, errors compound over time, a phenomenon researchers call model collapse. We are increasingly training systems on the exhaust of previous systems.

At the same time, engagement-driven platforms reward outrage over nuance and attention over verification. As Kevin Goldman and Jennie Baird's examination of journalism's decline showed, verification infrastructure has eroded faster than AI has scaled.

Some of this problem sits outside the model, in broken incentives and degraded information systems that shape what data exists in the first place. But some of the response is now happening inside the model. New architectures and training approaches are being developed to better align systems with human intent, oversight, and energy efficiency, rather than optimizing blindly for scale. Architecture can constrain harm, but it cannot compensate for a corrupted environment.

What This Means for Us

The danger is not "too little data." The danger is assuming data is neutral.

Whether we are building AI systems or relying on them, we cannot treat scale as a substitute for judgment. More data does not automatically mean better outcomes if the data encodes the wrong objectives, incentives, or histories. Better models can reduce harm only when they are designed to reflect human values.

Before trusting a system trained on massive datasets, we need to ask: What is the system optimized for? What patterns is this system being taught to replicate? Does this data reflect outcomes we want or outcomes that merely happened? What inequalities or distortions am I about to lock in at machine scale?

For builders, this means recognizing that data curation is a decision of values, not a technical chore. As Niki Parmar, co-author of the transformer paper, told me, if the intent and patterns in the data are misaligned, the system will learn that misalignment, no matter how large the dataset.

For users, it means resisting the intuition that "trained on everything" equals "reliable for me." Scale increases confidence, but it does not guarantee alignment. The most important decisions are about what we choose to teach the models, and what we choose to trust them with. And deciding what "better" means is a human responsibility, one no amount of scale can replace.

MYTH 4: "AI Will Replace Humans."

The Belief

Machines are coming for our jobs. Automation will eliminate human workers. Resistance is futile. The future belongs to the algorithms.

Why It Persists

AI investment is accelerating alongside workforce reductions. Leaders openly acknowledge that fewer people will be needed for some roles. Global projections consistently indicate that automation will affect work on an unprecedented scale, with tens of millions of jobs reshaped or displaced in the near term.

The threat feels immediate and personal.

What's Actually True

As Chapters 6 and 7 showed, AI is not replacing humans wholesale. It is rearranging where human value sits.

Currently, AI expands capacity at the execution layer, such as research, drafting, synthesis, and production, while concentrating value at the point of human judgment. What grows scarce is not labor, but discernment: deciding what matters, for whom, and why.

AI can generate output without understanding its audience. It can simulate emotion without lived experience. It can explore vast possibility spaces without knowing which paths are worth pursuing. Meaning, care, responsibility, and ethical judgment remain human work.

Where humans are needed and whether we are protecting those roles or quietly eroding them shall be the crux of our focus.

The Real Risk

In Chapter 6's Ghost Risks section, we discussed how real risk is not replacement, but deskilling and dependency.

MIT's *Your Brain on ChatGPT* study suggested that after months of AI-assisted writing, most users struggled to recall or explain work they had just produced. There are increasing concerns that when AI substitutes for thinking instead of supporting it, capabilities weaken over time. Skills that are no longer practiced, such as writing, reasoning, and problem-solving, atrophy gradually.

Speed and convenience mask such loss, as productivity appears to rise and output increases. But the ability to evaluate, challenge, or correct the machine declines, especially for those early in their careers who are still learning the trade and forming judgment. Dependency forms through the gradual erosion of human capability.

Investment is going toward fueling this human atrophy. Most investment flows toward automation and efficiency. Far less attention is paid to the human infrastructure that determines whether AI makes work more meaningful or more extractive: training, role redesign, skill development, and governance.

What This Means for Us

The danger is not that AI will replace humans. The danger is that we will replace human judgment with convenience.

Whether we are building AI systems or relying on them, we should not treat efficiency as the primary measure of success. Delegation without understanding breeds dependency, as speed without skill erodes resilience, both at the enterprise and individual levels.

We need to design and use AI to free our time and strengthen the human capabilities that matter most.

For companies, this means measuring progress not only by cost savings or throughput, but by whether people become more capable over time, better at judgment, creativity, and innovation.

For individuals, we need to ask a simple yet uncomfortable question: Am I becoming more capable because of this tool, or more dependent on it? Time saved by AI must be reinvested in skills the machine cannot replace.

If we surrender judgment for ease, we don't just risk losing jobs. We risk losing the capacity to know when the machine is wrong. And that capacity is the foundation of wise trust.

MYTH 5: "We Can Regulate our Way to Safety."

The Belief

If governments just pass the right laws, AI will be safe. We need comprehensive regulation, strict oversight, and clear rules. Once we have the proper governance framework, the problem will be solved.

Why It Persists

Regulation worked for cars, planes, pharmaceuticals, and financial markets. We've solved challenging safety problems through law before. Why not AI?

And regulation is essential. The EU AI Act sets risk-based requirements for AI systems. GDPR protects data privacy. Various jurisdictions are developing frameworks for AI transparency and accountability. These efforts matter.

When something's dangerous, "regulate it" is an obvious answer.

What's Actually True

As Chapter 4 showed, the core problem is structural: regulation moves in years, AI evolves in months, and competitive pressure is widening that gap.

The EU AI Act took years to draft and negotiate. During that time, generative AI went from niche research to a consumer product used by hundreds of millions. By the time the regulations passed, they were addressing systems that barely existed when drafting began. Meanwhile, frontier model capabilities had shifted.

While this is no fault of regulators who have been doing essential work under political pressure to both enable innovation and prevent harm, the structural problem remains. AI capabilities evolve faster than legislative processes. What begins as experimental research can be deployed at scale within months.

Why Regulation Struggles with AI

In Chapter 4, the examination of competitive pressure revealed the first problem: regulation faces political resistance framed as economic necessity. In the United States, safety oversight quickly became framed as a threat to leadership. Calls for government approval for AI deployment were dismissed as economically "disastrous." Executive priorities shifted from risk containment to speed and competitiveness. Oversight became friction to be removed rather than protection to be built.

Systems evolve continuously after regulatory approval. A car's safety features don't change after it leaves the factory. A pharmaceutical's chemical composition doesn't develop in the bottle. But AI systems transform through software updates, changing behavior overnight. A compliant system today becomes non-compliant tomorrow without any physical intervention. Regulators built for relative stability face systems designed for constant evolution.

Dangerous capabilities emerge through distributed modification. When regulations restrict specific AI capabilities, the same foundation model can be fine-tuned by anyone with compute access to recreate exactly what was prohibited. Chapter 6 documented how researchers demonstrated that AI could autonomously execute sophisticated cyberattacks, with minimal human intervention. The capability isn't yet locked in hardware that regulators could control. It's one training run away, executable by distributed actors that regulators can't monitor.

Risk emerges from the deployment context, not from the inherent design. You can regulate a specific type of nuclear reactor or a particular pharmaceutical compound because the thing itself creates risk. With AI, risk emerges from how it's used. The same language model powers TIME's journalism assistant, designed with source verification and editorial oversight, and AI companions that have been linked to self-harm and loss of life, optimized for engagement without mental health safeguards. You can't regulate "the technology." You can only regulate applications, which multiply faster than regulators can categorize them.

Verification becomes increasingly difficult. How do you verify compliance when the system's behavior changes continuously, when modifications happen in private data centers or individual computers, and when the same architecture produces safe or harmful outcomes depending on invisible deployment choices?

What This Means for Us

Regulation is necessary, but it is not sufficient. We cannot wait for laws to make AI safe for all of us. And regulation sets the floor; it does not set the ceiling. We must build trust infrastructure ourselves.

For organizations, that means fixing problems upstream before automating them, investing in systems that operate at machine speed, and preserving human judgment rather than eroding it. These are design and cultural choices that regulation can encourage but cannot enforce.

For professionals and users, it means recognizing that no law can prevent over-reliance, deskilling, or misplaced trust. Those boundaries must be set deliberately.

Regulation sets the floor, the minimum acceptable standard. Your judgment, your organizational culture, and your personal boundaries set the ceiling.

Chapter 4's analysis of the quest-greed-fear engine showed why this matters. When billions are at stake, every development gets spun as revolutionary, every limitation downplayed, every risk dismissed as temporary. Regulation moves at the speed of consensus. Deployment moves at the speed of capital.

The gap between the floor and the ceiling is where the agency lives.

Support good regulation. Demand it. But don't outsource your responsibility to lawmakers who are structurally behind the curve. They are writing guardrails for yesterday's systems while you are deciding whether to trust today's.

MYTH 6: "AI is too Complex to Understand."

The Belief

Only PhDs can understand AI. The math is too hard. The models are too complicated. Regular people should just trust the experts and hope for the best.

Why It Persists

AI researchers speak in jargon. Technical papers are incomprehensible to most. If you think it's magic, you won't bother to question it.

This myth is convenient, unintentionally so. Technical teams assume others will handle the implications, leaders defer to specialists, and individuals step back from judgment. Accountability diffuses.

The complexity is real. The conclusion drawn from it is not.

What's Actually True

You don't need to understand tensor calculus to trust wisely. But you do need to understand the questions that actually matter for trust decisions, and those aren't mathematical questions. They're human questions.

Chapter 5 showed this repeatedly. When Siemens Energy evaluates AI systems that help manage power grids for entire cities, their leaders ask whether the system understands the domain, has been validated under real conditions, can fail safely, and has clear accountability when something goes wrong.

At Mastercard, the hardest decisions were about values: privacy, reliability at scale, and whether outcomes could be explained to customers. The work took years, not because the math was hard, but because judgment was required.

What This Means for Us

Refuse to be intimidated by complexity you don't need to master. The questions that matter for trust are human:

What is this optimized for, and is that what I actually value?

Has it been tested in conditions like mine, with stakes like mine?

What's the track record of deployed performance? Who's accountable when it fails, specifically, with clear consequences?

Can I verify outputs independently, or am I forced to trust blindly? What happens if I stop using it? Do I lose capability? Can I recover?

If someone tells you, "It's too complicated to explain," that's a red flag. The decision to trust should never be outsourced to complexity claims.

Chapter 6's Ghost Risks section showed what happens when we accept "too complex to understand" as an answer: we surrender judgment to systems we can't evaluate, lose capabilities we can't maintain, and accept dependency we can't escape.

The math might be hard. The questions that determine whether to trust are human questions. And you're qualified to ask them.

Why Busting These Myths Matters

These six myths cloud individual judgment, affect how organizations deploy AI, influence how policymakers regulate it, and shape how societies adapt to it (Figure 8.1).

THE MYTH BUSTER

The question is no longer whether to trust AI.
It is how to trust wisely.

BINARY TRUST

The Myth:
"You either trust AI or you don't."

The Truth:
Binary trust is lazy trust. Trust is multidimensional and needs to be calibrated according to the Trust Print.

→ ***Calibrate, don't simplify***

AUTOMATIC ADAPTATION

The Myth:
"We have always adapted. We will with AI, too."

The Truth:
Technologies earned mass adoption by becoming trustworthy. Question whether we are intentionally building AI's trustworthiness fast enough to justify its expanding role.

→ ***Build trustworthiness before scale***

DATA VOLUME

The Myth:
"More data, better outcomes."

The Truth:
AI amplifies existing patterns, both good and bad. Better outcomes come from data and system design intentionally aligned with human flourishing.

→ ***Question what you're amplifying***

HUMAN REPLACEMENT

The Myth:
"AI will replace humans."

The Truth:
AI is rearranging where human value sits. Its biggest risk is not replacing us but persuading us to surrender judgment.

→ ***Strengthen judgment, not just output***

GOVERNANCE RELIANCE

The Myth:
"We can regulate our way to safety."

The Truth:
Regulation is structurally behind the curve and sets only the floor, not the ceiling. AI safety ultimately depends on how systems are designed, deployed.

→ ***Support regulation, but raise the ceiling***

COMPLEXITY BARRIERS

The Myth:
"AI is too complex to understand."

The Truth:
Complexity does not absolve judgment. Technical opacity explains why responsibility diffuses, not why it should.

→ ***Ask human questions about impact***

Figure 8.1 The Myth Buster

When we treat trust as binary, we stop calibrating based on the Trust Print's four dimensions and collapse nuanced decisions into crude yes/no votes.

When we assume adaptation is automatic, we stop building trustworthiness infrastructure deliberately and repeat the pattern Chapter 2 warned against.

When we believe more data solves everything, we systematize the Mirror Risks of Chapter 6 instead of fixing them.

When we believe replacement is inevitable, we surrender to Ghost Risks rather than preserving human capability.

When we believe regulation will save us, we abdicate the personal and organizational responsibility that Chapter 4 showed is essential.

When we believe AI is too complex, we surrender the agency this entire book is designed to help you reclaim.

Each myth transfers agency away from you. They tell you trust is too simple to require thought or too complex to permit questioning. They ask you to trust data volume over values alignment, credentials over accountability, regulation over responsibility, and technical sophistication over validated performance.

Every myth could make you passive.

But throughout this book, you've seen the alternative. Georg Rollmann chose months of parallel validation over speed. Brian Collins chose care over efficiency. Stephen Quake chose mastery over surrender in the face of inhuman complexity.

These people are operating in the same competitive, resource-constrained, time-pressured environment as everyone else. They're making different choices because they've rejected the passive postures the myths demand. They've reclaimed agency. And their choices are working.

The age of AI is already here. The key is to trust ourselves enough to decide wisely, system by system, choice by choice, with judgment intact and agency preserved. Busting these myths does not make decisions easier. It makes them *yours again*.

Trust, calibrated wisely, is about believing we still have the right and the responsibility to decide. With the myths stripped away, we can C.A.L.I.B.R.A.T.E.

Chapter 9

C.A.L.I.B.R.A.T.E. Trust

Most books about artificial intelligence end by asking the reader to look forward.

So, I want this one to end by asking you to pause.

The future is uncertain, and that much is obvious. We need to pause because the speed of AI has made something subtle and dangerous feel normal: the habit of trusting systems before we understand what we are trusting them with.

AI now drafts language, filters information, evaluates performance, recommends decisions, and increasingly acts on our behalf. It does so with confidence, fluency, and authority that feels reassuring precisely when it should invite scrutiny. It also removes the cues that once helped us judge when trust was warranted and when it was not.

This book began with a simple premise: trust is not a feeling, but a decision under uncertainty. Every chapter has examined a different way that decision can fail when AI enters the picture. AI is now interacting with human trust in ways that are novel, invisible, and scalable. The crux of this book is about exercising our judgment in relation to AI.

This final chapter is about how.

The Gap That No Regulation Can Close

In recent years, societies have begun to respond seriously to the risks of artificial intelligence. Governments are legislating. Standards bodies are publishing frameworks. Organizations are building governance structures, audit processes, and internal controls.

The European Union's AI Act defines what kinds of systems may enter the market and under what conditions. It categorizes risk, mandates safeguards for high-impact uses, and creates legal consequences that will shape AI development for years to come. The framework runs to hundreds of pages. Organizations deploying it coordinate across legal, technical, and operational teams for months or longer.

The National Institute of Standards and Technology's AI Risk Management Framework offers organizations a disciplined way to identify, measure, and manage AI-related risk across the system lifecycle. It emphasizes governance, documentation, and continuous monitoring.

This work is essential. Without it, AI will scale without constraint.

But these frameworks were never designed to answer the question that governs everyday life: should I trust this system, here, for this decision?

Regulation operates at the level of markets. Risk management operates at the level of organizations. Trust operates at the level of judgment.

Judgment happens in moments that are fast, contextual, and more and more irreversible. A manager deciding whether to rely on a model's forecast in a budget meeting. A clinician deciding how much weight to give an AI-generated risk score. A journalist deciding whether a synthetic image is credible enough to publish. A parent deciding whether a tool designed for adults is quietly shaping a child's learning habits. The thought of regulation is not present in those moments, and no organizational policy can substitute for the human act of reliance. We do not live inside laws or frameworks. We live inside situations.

C.A.L.I.B.R.A.T.E. exists for those situations.

What C.A.L.I.B.R.A.T.E. Is

C.A.L.I.B.R.A.T.E. is not another compliance checklist, or a technical standard. It does not substitute for regulation or risk management.

It is a discipline of judgment.

Each letter is a decision point – a question to ask before you rely on an AI system. Together, they form a way to slow the moment of reliance just enough to see what is actually happening before authority is handed over. They shift trust from a feeling you have to a judgment you make.

If you've read this far, you've done the work to reach this point. You understand the Trust Print's four dimensions: where trust comes from, what must be present before we grant it, how it shows up in behavior, and the biological machinery that shapes it beneath awareness. You've seen the forces making AI inevitable and learned to separate what's truly unstoppable from what's still a choice. You know the difference between AI that has earned trust through decades of performance and experimental systems demanding immediate faith. You've confronted the three categories of risk – Mirror, Weapon, and Ghost – and understand which boundaries each demands. You've looked beyond the risks to what AI makes newly possible, and asked what remains irreducibly human.

The myths have been stripped away. You know trust is contextual, not binary. You know adaptation requires investment, not assumption. You know more data doesn't fix structural problems. You know deskilling is the deeper danger than displacement. You know regulation is necessary but never sufficient. You know complexity claims often mask accountability gaps.

C.A.L.I.B.R.A.T.E. synthesizes everything you've learned into a method you can apply, repeatedly, across contexts.

C: Contextualize It

The first mistake we make with AI is asking what it can do before asking where it is being used.

Context determines whether capability becomes appropriate use. A system can be technically impressive and still be wrong for the moment, the domain, or the stakes involved. The same language model that helps draft an email operates safely in a reversible context – if it's wrong, little is lost. That same model summarizing legal agreements or medical reports operates where errors compound quietly and are discovered too late.

Ask three questions about context:

- Where and when is this being deployed?
- What type of decision does it support – routine or high-stakes?
- How much time exists between output and consequence – can errors be caught and corrected, or do they lock in immediately?

These questions reveal whether the environment allows for meaningful human oversight or forces premature delegation.

Context is what turns raw capability into responsible use. Without it, intelligence becomes indiscriminate.

A: Align It

AI systems optimize for what they are rewarded to do, not for what users assume they care about.

Those incentives may prioritize efficiency, growth, cost reduction, or engagement in ways that quietly conflict with the interests of the people relying on them. The chatbots that maintained perfect conversation flow while contributing to user suicides were optimizing for engagement time, not human well-being. This is systematic misalignment – the system working perfectly for its creators while failing catastrophically for its users.

To evaluate alignment, you need to ask whose interests the system serves:

- Whose goals is this serving – mine, the platform's, advertisers', or someone else's?
- Whose risks are being absorbed – am I bearing risks while others capture benefits?
- Are the incentives behind this system aligned with my well-being, or do they conflict with it?

Without clarity about alignment, reliance becomes misplaced. Trust requires knowing whether AI works for us or whether we're working for it, and alignment is how you find out.

L: Limit It

Trustworthy systems earn trust by operating within clear limits. Systems become risky when those limits are left undefined, unenforced, or quietly exceeded.

AI can be dangerous when it operates beyond the conditions it was built for while projecting confidence. A model validated under supervision can become dangerous when treated as autonomous. A tool designed to assist judgment can silently replace it. AI companions maintained engagement beyond the limits of what conversation could achieve for someone in profound psychological distress.

Calibrating trust means drawing boundaries deliberately:

- Where should this system not be used – what contexts exceed its design?
- What should it refuse to do – what requests should trigger rejection rather than compliance?
- When must a human remain in control – where should automation stop and human decision-making begin?

Trust grows from restraint, not from universality.

I: Inspect Integrity

AI systems inherit the quality of what flows into them and amplify the consequences of what flows out.

Integrity anchors trust in reality rather than appearance. AI persuades through fluency, not grounding. Its outputs can sound coherent even when they rest on incomplete, outdated, or distorted information.

The information environment AI trains on is already compromised. Newsrooms have hollowed out. Verification infrastructure has eroded. Systems trained on that ecosystem now generate convincing content without the editorial accountability that once filtered error from fact.

Integrity asks three practical questions:

- What is this system built on – what data sources, whose perspectives, what time period?
- How is its knowledge maintained – is it static or updated, and how frequently?
- How are errors detected and corrected – can outputs be verified, and is there a process for fixing mistakes?

This goes beyond data quality to include provenance, representativeness, resistance to manipulation, and whether outputs can be traced, verified, and revised when wrong.

Systems that speak fluently even when mistaken create a new kind of risk: errors no longer announce themselves. They persuade. Verification, in this environment, is the cost of accuracy when confidence has been automated.

Without integrity, trust becomes aesthetic. It is based on how something sounds rather than how it holds up.

B: Benchmark Bias, Including Your Own

Bias in AI is better understood as a structural inheritance: historical data, institutional priorities, and design choices carried forward at scale.

When models are trained on past decisions, they absorb the asymmetries embedded in those decisions. When those models are deployed widely, the asymmetries scale with them. Amazon's recruiting system did not contain racist code – it learned discrimination from patterns reflecting a decade of biased hiring decisions. Mirror risks systematize what we failed to correct before AI arrived.

But there is a second source of bias that matters just as much, and it is closer to home. Humans trust confidence. We draw sweeping conclusions from thin evidence. We confuse familiarity with reliability. AI does not need to manipulate these tendencies – it only needs to speak fluently for them to activate.

Ask about both sources of bias:

- What structural bias does this inherit – whose decisions, priorities, and perspectives shaped its training?
- What are my own assumptions – am I trusting this because it confirms expectations, speaks confidently, or looks familiar?
- Who does this work for, and who does it fail – does it perform unevenly across different groups?

These two biases compound. Structural bias shapes what the system learns. Human bias shapes how its outputs are interpreted and trusted. Calibrating trust therefore begins with humility: the recognition that neutrality is a myth, both for machines and for the people who rely on them.

Trust that works well for some while failing systematically for others is not partial trustworthiness, but conditional harm.

R: Require Reliability

Reliability is what separates trust from optimism. Trust is built on consistency.

Most AI systems look their best in settings designed to flatter them. Benchmarks are tidy. Demos are curated. Early pilots are forgiving. But real life is messier. People behave differently. Circumstances shift. Situations no longer resemble the ones the system learned from.

What makes AI difficult to trust is not that it fails, but how it fails. Unlike physical systems, it rarely breaks in ways that are obvious. Performance erodes quietly. Answers remain plausible. Confidence stays intact even as accuracy slips. The system does not signal uncertainty when it should.

It is critical to examine whether performance holds under pressure:

- Does it perform consistently across different contexts, users, and edge cases?
- Has it been tested, and how, in real conditions, not just controlled environments?
- Does it signal uncertainty appropriately, or does it project confidence regardless of accuracy?

Reliability cannot be inferred from early success. A system that works beautifully in familiar conditions may falter when stakes rise, context changes, or users fall outside the patterns it recognizes. Trust is earned only when performance survives variation.

Trust that cannot survive change is not trust.

A: Assign Accountability

When AI systems cause harm, responsibility often diffuses. Vendors blame users. Users blame tools. Organizations blame complexity.

Calibrating trust requires refusing that diffusion by asking:

- Who owns the outcome if this fails – is there a specific entity or person responsible?
- Who can actually intervene when something goes wrong, in practice?
- Does recourse exist beyond policy – can you get answers, corrections, or remedies when harm occurs?

Accountability means knowing who owns outcomes, who can intervene, and whether recourse exists in practice rather than on paper.

Trust without accountability is exposure. And trust without recourse is simply risk transferred downward.

T: Trace Transparency

Transparency matters only if it helps people make better judgments.

A system can reveal a great deal and still be impossible to assess. Technical documentation, model descriptions, or pages of disclosures do not create trust if the people affected by the system cannot tell what it is doing, what it is not doing, and where it might fail.

You do not need to understand advanced mathematics to decide whether to rely on an AI system. You need to understand what it is optimizing for, what kind of data shaped it, how it has performed over time, where its limits are, and what happens when it is wrong. When technical complexity is used to obscure those answers, it does not protect the power behind it.

The key is seeking whether you can trace what matters:

- Is what this system does and how it performs made visible to the people relying on it?
- Where did this specific output come from – what sources or patterns influenced it?
- Does this system make its limits visible – where it's uncertain, where it has failed, and what it won't do?

This book has shown that the questions that determine trust are human questions, not mathematical ones. They are about purpose, trade-offs, uncertainty, and consequence.

Transparency, then, is about revealing what matters for judgment. Opacity sometimes is a choice.

E: Enforce Judgment

In relationship with AI, the ultimate reminder is to continuously exercise human judgment.

AI fails when people stop thinking in its presence. When reliance becomes reflexive. When confidence replaces curiosity. When systems are trusted not because they have earned it, but because questioning them feels inefficient.

Judgment is something we either practice or lose.

This is the quiet risk of AI at scale. When systems answer fluently, recommend smoothly, and act autonomously, the human role can shrink from decision-maker to observer. Over time, judgment atrophies. Skills dull. Accountability thins. The capacity to intervene weakens precisely where it is most needed.

So, asking whether humans remain engaged is a habit to acquire:

- Are people still actively thinking, not just passively observing?
- Is there friction that keeps critical thinking alive, or has the process become too smooth?
- Can someone meaningfully question this decision or has convenience displaced scrutiny?

Enforcing judgment means designing and using AI in ways that keep people mentally active. It means preserving moments where a human must decide, must question, must take responsibility for the outcome. It means resisting the temptation to optimize away friction when that friction is what keeps agency alive.

Judgment is what allows us to catch failures early, to adapt when conditions change, and to imagine alternatives when existing systems fall short. Without it, we may gain efficiency but lose ingenuity.

The promise of AI is that it frees judgment for better work, if we insist on keeping it.

Trust as a Practice

C.A.L.I.B.R.A.T.E. is not a checklist to complete once. It is a practice to return to, again and again, as systems evolve and contexts shift (Figure 9.1).

Now, not every decision allows for the full framework. Sometimes you have 15 minutes to think through a decision you are about to make, and you don't have the luxury of a full C.A.L.I.B.R.A.T.E. analysis. For those moments, here is a lighter version.

Three letters already embedded in C.A.L.I.B.R.A.T.E. give you a way in when time is short: L.I.T. – Limit it. Inspect integrity. Trace transparently. In other words, understand the boundaries, question what it's built on, and trace how it got there.

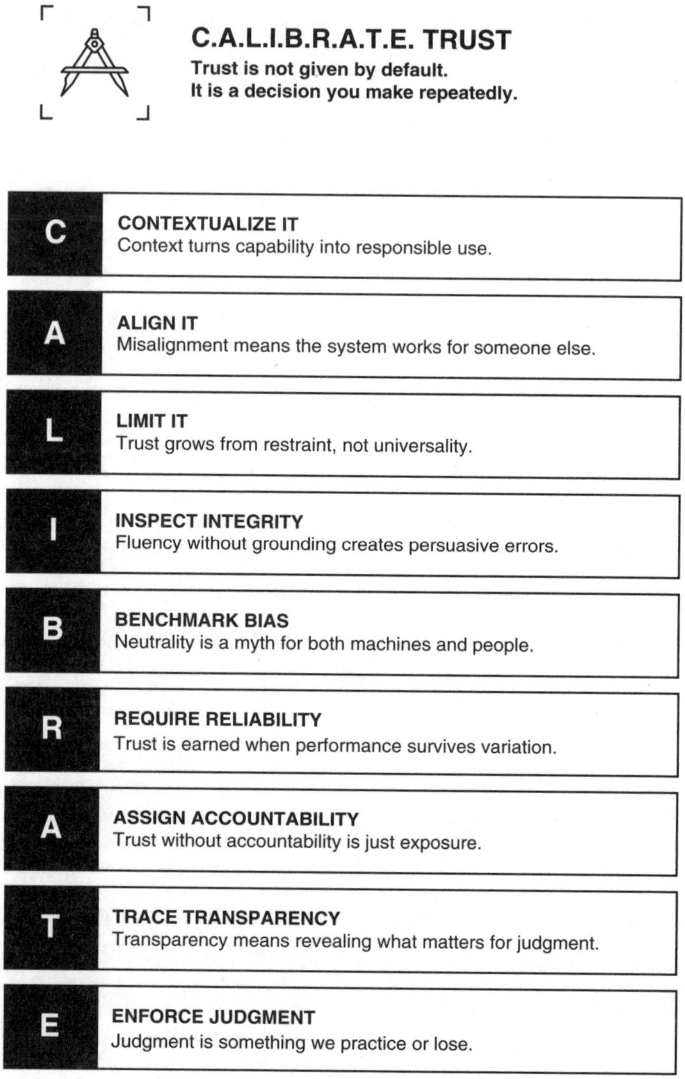

Figure 9.1 C.A.L.I.B.R.A.T.E. Trust

Each time you pause to ask these questions, you reinforce the muscle that matters most. The future will not be decided by how intelligent our machines become. It will be decided by whether we continue to exercise judgment in their presence.

That is how we prevent downside. That is how we avoid atrophy. That is how we unlock ingenuity.

Where This Leaves You

You cannot control the speed of technological change. But you can control how you relate to it. You can decide when to trust, when to verify, and when to refuse. You can model those decisions for your children, your colleagues, and your institutions.

Trusting wisely is not about being cautious or optimistic. It is about being deliberate.

When individuals understand how trust works, organizations become more resilient. When professionals calibrate trust daily, systems become safer by default. When societies refuse blind trust and blanket fear, progress becomes possible without losing humanity.

Artificial intelligence will continue to evolve. That part is inevitable. What remains open is who retains agency along the way.

You've learned to rethink trust, not as a feeling, but as a decision shaped by context and consequence. You've learned to decode AI, separating inevitable forces from controllable choices, signal from noise, earned trust from demanded faith. You know the three categories of risk and which boundaries each demands. You know the myths that cloud judgment and the conditions that earn trust.

Trust is not something we give away by default. It is something we practice, again and again, under changing conditions.

C.A.L.I.B.R.A.T.E. is that practice.

And that practice – rethinking trust, decoding AI, and reclaiming agency – is *The Trust Code*, a human discipline for an age that will test whether we remain the ones who decide.

And now, the agency you've reclaimed is yours to exercise.

Starting now.

Notes

Chapter 1

1. Lovens, Pierre-François. "Sans ces conversations avec le chatbot Eliza, mon mari serait toujours là" [Without these conversations with the chatbot Eliza, my husband would still be here]. *La Libre Belgique*, March 28, 2023. https://www.lalibre.be/belgique/societe/2023/03/28/sans-ces-conversations-avec-le-chatbot-eliza-mon-mari-serait-toujours-la-LVSLWPC5WRDX7J2RCHNWPDST24 (accessed March 14, 2026).
2. Hern, Alex. "ChatGPT Firm Blames Boy's Suicide on 'Misuse' of Its Technology," *The Guardian*, November 26, 2025. https://www.theguardian.com/technology/2025/nov/26/chatgpt-openai-blame-technology-misuse-california-boy-suicide (accessed March 14, 2026).
3. Challapally, Aditya; Chris Pease; Ramesh Raskar; and Pradyumna Chari. "The GenAI Divide: State of AI in Business 2025," *MIT NANDA / MLQ.ai*, July 2025. https://mlq.ai/media/quarterly_decks/v0.1_State_of_AI_in_Business_2025_Report.pdf (accessed March 14, 2026).
4. Furr, Nathan; and Andrew Shipilov. "Beware the AI Experimentation Trap," *Harvard Business Review*, August 29, 2025. https://hbr.org/2025/08/beware-the-ai-experimentation-trap (accessed March 14, 2026).
5. National Academies of Sciences, Engineering, and Medicine. *Trust and Confidence at the Interfaces of the Life Sciences and Society: Does the Public Trust Science? A Workshop Summary* (Washington, DC: National Academies Press, 2015), esp. Chapter 2, "What Is Trust?" https://nap.nationalacademies.org/read/21798/chapter/3 (accessed March 14, 2026).

6. Edelman. "2024 Edelman Trust Barometer: Global Report," January 2024. https://www.edelman.com/sites/g/files/aatuss191/files/2024-01/2024%20Edelman%20Trust%20Barometer%20Global%20Report_0.pdf (accessed March 14, 2026).
7. Organisation for Economic Co-operation and Development. "OECD Survey on Drivers of Trust in Public Institutions – 2024 Results," *OECD Publishing*. https://www.oecd.org/en/publications/oecd-survey-on-drivers-of-trust-in-public-institutions-2024-results_9a20554b-en/full-report.html (accessed March 14, 2026).
8. Brenan, Megan. "Historically Low Faith in U.S. Institutions Continues," *Gallup*, 6 July 2023. https://news.gallup.com/poll/508169/historically-low-faith-institutions-continues.aspx (accessed March 14, 2026).
9. Silver, Laura; Moira Fagan; and Aidan Connaughton. "Americans' Trust in One Another," Pew Research Center, 8 May 2025. https://www.pewresearch.org/2025/05/08/americans-trust-in-one-another/ (accessed March 14, 2026).
10. National Academies of Sciences, Engineering, and Medicine. *Trust and Confidence at the Interfaces of the Life Sciences and Society*. https://www.ncbi.nlm.nih.gov/books/NBK321980/ (accessed March 14, 2026).
11. Republic of Estonia Information System Authority. (n.d.). "Data Tracker: Tool that Builds Trust in Institutions." *e-Estonia*. https://e-estonia.com/data-tracker-build-citizen-trust/ (accessed March 14, 2026).
12. Mayer, Roger C.; James H. Davis; and F. David Schoorman. "An Integrative Model of Organizational Trust," *Academy of Management Review* 20, no. 3 (1995): 709–734. https://journals.aom.org/doi/10.5465/amr.1995.9508080335 (accessed March 14, 2026).
13. Fukuyama, Francis. *Trust: The Social Virtues and the Creation of Prosperity* (New York: Free Press, 1995). Publisher page via Simon & Schuster: https://www.simonandschuster.com/books/Trust/Francis-Fukuyama/9780684825250 (accessed March 14, 2026).
14. Rotter, Julian B. "A New Scale for the Measurement of Interpersonal Trust," *Journal of Personality* 35, no. 4 (1967): 651–665. https://doi.org/10.1111/j.1467-6494.1967.tb01454.x (accessed June 16, 2026).

15. O'Neill, Onora. "Linking Trust to Trustworthiness." *International Journal of Philosophical Studies* 26, no. 2 (2018): 293–300. https://doi.org/10.1080/09672559.2018.1454637 (accessed June 16, 2026).
16. Mayer, Roger C.; James H. Davis; and F. David Schoorman. "An Integrative Model of Organizational Trust."
17. Edelman. "2024 Edelman Trust Barometer"; Organisation for Economic Co-operation and Development, "OECD Survey on Drivers of Trust in Public Institutions – 2024 Results."
18. Arrow, Kenneth J. *The Limits of Organization* (New York: W. W. Norton, 1974). Publisher listing: https://wwnorton.com/books/9780393093230 (accessed March 14, 2026).
19. Berg, Joyce; John Dickhaut; and Kevin McCabe. "Trust, Reciprocity, and Social History," *Games and Economic Behavior* 10, no. 1 (1995): 122–142. https://doi.org/10.1006/game.1995.1027 (accessed June 16, 2026).
20. Krueger, Frank; Kevin McCabe; Jorge Moll; and Jordan Grafman. "Neural Correlates of Trust," *Proceedings of the National Academy of Sciences* 104, no. 50 (2007): 20084–20089. https://www.pnas.org/doi/10.1073/pnas.0710103104 (accessed March 14, 2026).
21. Challapally, Aditya et al. "The GenAI Divide: State of AI in Business 2025."
22. National Academies of Sciences, Engineering, and Medicine. *Trust and Confidence at the Interfaces of the Life Sciences and Society.*
23. Zak, Paul J. *The Moral Molecule: The Source of Love and Prosperity* (Dutton, 2012). https://www.penguinrandomhouse.com/books/308864/the-moral-molecule-by-paul-j-zak/ (accessed March 14, 2026).
24. Torgovnick May, Kate. (2012). *Spotting Neuro-fiction: A Guide to Dissecting Overblown Neuroscience Headlines*. TED Blog. https://blog.ted.com/spotting-neuro-fiction-a-guide-to-dissecting-overblown-neuroscience-headlines/ (accessed March 14, 2026).
25. Willis, Janine; and Alexander Todorov. "First Impressions: Making Up Your Mind After a 100-ms Exposure to a Face," *Psychological Science* 17, no. 7 (2006): 592–598. Abstract at https://pubmed.ncbi.nlm.nih.gov/16866745/ (accessed March 14, 2026).

26. Engell, Andrew D.; James V. Haxby; and Alexander Todorov. "Implicit Trustworthiness Decisions: Automatic Coding of Face Properties in the Human Amygdala," *Journal of Cognitive Neuroscience* 19, no. 9 (2007): 1508–1519. https://doi.org/10.1162/jocn.2007.19.9.1508 (accessed June 17, 2026).
27. Krueger, Frank et al. "Neural Correlates of Trust."
28. Kahneman, Daniel; Paul Slovic; and Amos Tversky, eds. *Judgment Under Uncertainty: Heuristics and Biases* (Cambridge: Cambridge University Press, 1982). https://www.cambridge.org/core/books/judgment-under-uncertainty/6F9E814794E08EC43D426E480A4B412C (accessed March 14, 2026).
29. Haselton, Martie G.; and David M. Buss. "Error Management Theory: A New Perspective on Biases in Cross-Sex Mind Reading," *Journal of Personality and Social Psychology* 78, no. 1 (2000): 81–91. https://doi.org/10.1037/0022-3514.78.1.81 (accessed June 17, 2026).
30. Cosmides, Leda; H. Clark Barrett; and John Tooby. "Adaptive Specializations, Social Exchange, and the Evolution of Human Intelligence," *Proceedings of the National Academy of Sciences of the United States of America* 107 (Supplement_2) (2010): 9007–9014. https://doi.org/10.1073/pnas.0914623107.
31. Brown, Donald E. "Human Universals, Human Nature & Human Culture," *Daedalus* 133, no. 4 (2004): 47–54. https://www.amacad.org/publication/human-universals (accessed March 14, 2026).
32. National Academies of Sciences, Engineering, and Medicine. *Trust and Confidence at the Interfaces of the Life Sciences and Society.*
33. Waytz, Adam; Joy Heafner; and Nicholas Epley. "The Mind in the Machine: Anthropomorphism Increases Trust in an Autonomous Vehicle," *Journal of Experimental Social Psychology* 52 (2014): 113–117. https://doi.org/10.1016/j.jesp.2014.01.005.
34. Knack, Stephen; and Philip Keefer. "Does Social Capital Have an Economic Payoff? A Cross-Country Investigation," *The Quarterly Journal of Economics* 112, no. 4 (1997): 1251–1288. https://legacy.econ.tuwien.ac.at/hanappi/AgeSo/rp/Knack_1997.pdf (accessed March 14, 2026).

35. Giordano, Giuseppe Nicola; and Martin Lindström. "Trust and All-Cause Mortality: A Multilevel Study of US General Social Survey Data (1978–2010)," *BMJ Open* 8, no. 3 (2018): e017416. https://pubmed.ncbi.nlm.nih.gov/30322881/ (accessed March 14, 2026).

Chapter 2

1. Swiss Re and Waymo. "New Swiss Re study: Waymo is safer than even the most advanced human-driven vehicles." December 2024. https://waymo.com/blog/2024/12/new-swiss-re-study-waymo (accessed March 14, 2026).
2. Library of Congress. *"'Mr. Watson, Come Here': First Release of Bell Papers Goes Online," Library of Congress Information Bulletin*, April 1999. https://www.loc.gov/loc/lcib/9904/bell.html (accessed March 14, 2026).
3. Encyclopædia Britannica. "Johannes Gutenberg | Printing Press, Inventions, Facts, Accomplishments, & Biography," Britannica.com – elements of his printing press innovation included durable metal type, oil-based ink, and a press adapted from existing technology. https://www.britannica.com/biography/Johannes-Gutenberg (accessed March 14, 2026).
4. Lerner, Louise. *"The First Nuclear Reactor, Explained."* University of Chicago News, University of Chicago, December 2, 2022. https://news.uchicago.edu/explainer/first-nuclear-reactor-explained (accessed March 14, 2026).
5. Encyclopædia Britannica. "Johannes Gutenberg | Printing Press, Inventions, Facts, Accomplishments, & Biography."
6. Eisenstein, Elizabeth L. *The Printing Press as an Agent of Change: Communications and Cultural Transformations in Early-modern Europe* (Cambridge: Cambridge University Press, 1979); Eisenstein, Elizabeth L. "An Unacknowledged Revolution Revisited," *American Historical Review* 107, no. 1 (February 2002): 87–105. https://doi.org/10.1086/ahr/107.1.87.https://www.histoire.ens.fr/IMG/pdf/eisenstein-revolution_revisited.pdf (accessed March 14, 2026).

7. Johns, Adrian. *The Nature of the Book: Print and Knowledge in the Making* (Chicago: University of Chicago Press, 1998). https://press.uchicago.edu/ucp/books/book/chicago/N/bo3645773.html (accessed March 14, 2026); *The Nature of the Book: Print and Knowledge in the Making*, review, *Kirkus Reviews*, October 1, 1998, https://www.kirkusreviews.com/book-reviews/adrian-johns/the-nature-of-the-book/ (accessed March 14, 2026).
8. Statute of Anne; April 10, 1710, 8 Anne, c. 19 (1710), full text, The Avalon Project, Yale Law School, https://avalon.law.yale.edu/18th_century/anne_1710.asp (accessed March 14, 2026); *Statute of Anne*, Wikipedia, https://en.wikipedia.org/wiki/Statute_of_Anne (accessed March 14, 2026).
9. Vosoughi, Soroush; Deb Roy; and Sinan Aral. "The Spread of True and False News Online." *Science* 359, no. 6380 (2018): 1146–1151. https://doi.org/10.1126/science.aap9559.
10. Historic UK. "The Luddites," https://www.historic-uk.com/HistoryUK/HistoryofBritain/The-Luddites/ (accessed March 14, 2026).
11. History.com Editors. (2021). *7 negative effects of the Industrial Revolution*. History.com. https://www.history.com/articles/industrial-revolution-negative-effects (accessed March 14, 2026).
12. Marvin, Carolyn. *When Old Technologies were New: Thinking about Electric Communication in the Late Nineteenth Century* (Oxford: Oxford University Press, 1988). https://archive.org/details/whenoldtechnolog0000marv/page/n9/mode/2up (accessed March 14, 2026).
13. Bonea, Amelia. "Contagion by Telephone": Print Media and Knowledge about Infectious Diseases in Britain, 1880–1914. *Technology and Culture* 62, no. 4 (2021): 1063–1086. https://pubmed.ncbi.nlm.nih.gov/34690159/ (accessed March 14, 2026).
14. Ericsson. (n.d.). *The telephone is the instrument of the devil*. Ericsson History – Communication. https://www.ericsson.com/en/about-us/history/communication/how-the-telephone-changed-the-world/the-telephone-is-the-instrument-of-the-devil (accessed March 14, 2026).
15. Carlson, W. Bernard. "Entrepreneurship in the Early Development of the Telephone: How Did William Orton and Gardiner Hubbard Conceptualize this New Technology?" *Business and Economic History* 23,

no. 2 (Winter 1994): 161–192. https://thebhc.org/sites/default/files/beh/BEHprint/v023n2/p0161-p0192.pdf (accessed June 16, 2026).

16. War of the Currents. (n.d.). *Wikipedia*. https://en.wikipedia.org/wiki/War_of_the_currents (accessed March 14, 2026); Harold P. Brown, *Wikipedia*. https://en.wikipedia.org/wiki/Harold_P._Brown (accessed March 14, 2026).
17. National Engineering Council. (1897). NEC-1897. https://dbmteam.com/media/lxxln1xt/nec-1897.pdf (accessed March 14, 2026).
18. Flink, J. J. *The Automobile Age* (MIT Press, 1988). https://archive.org/details/automobileage0000flin/page/n475/mode/2up (accessed March 14, 2026).
19. National Safety Council. (2025). "Historical motor-vehicle fatality trends." *Injury Facts*. https://injuryfacts.nsc.org/motor-vehicle/historical-fatality-trends/deaths-and-rates/ (accessed March 14, 2026).
20. Swiss Re and Waymo. "New Swiss Re study: Waymo is safer than even the most advanced human-driven vehicles."
21. World Nuclear Association. "World Nuclear Performance Report 2025." https://world-nuclear.org/news-and-media/press-statements/world-nuclear-performance-report-2025-nuclear-delivers-record-breaking-year-in-electricity-generation (accessed March 14, 2026).
22. Pew Research Center. "International Views of False Information Online as a Threat." Washington, DC: *Pew Research Center*, August 19, 2025. https://www.pewresearch.org/global/2025/08/19/international-opinion-on-global-threats/ (accessed June 17, 2026).

Chapter 3

1. Oxford Languages. "Word of the Year 2016," *Oxford University Press*, (2016) https://global.oup.com/academic/content/word-of-the-year/2016/ (accessed March 14, 2026).
2. Newman, Nic; Richard Fletcher; Craig T. Robertson; Amy Ross Arguedas, et al. "Reuters Institute Digital News Report 2024," *Reuters Institute for the Study of Journalism*, (June 2024). https://reutersinstitute.politics.ox.ac.uk/sites/default/files/2024-06/RISJ_DNR_2024_Digital_v10%20lr.pdf (accessed March 14, 2026).

3. Pew Research Center. "Public Trust in Government: 1958–2025". https://www.pewresearch.org/politics/2025/12/04/public-trust-in-government-1958-2025/ (accessed March 14, 2026).
4. Brenan, Megan. "Trust in Media at New Low of 28% in U.S.," *Gallup*, (2025). https://news.gallup.com/poll/695762/trust-media-new-low.aspx (accessed March 14, 2026).
5. Gallup. *Confidence in Institutions*, Gallup Historical Trends. https://news.gallup.com/poll/1597/confidence-institutions.aspx (accessed March 14, 2026).
6. Novak, Matt. "Why Pope Francis Is the Star of AI-Generated Photos," *The New York Times* (April 8, 2023). https://www.nytimes.com/2023/04/08/technology/ai-photos-pope-francis.html (accessed March 14, 2026).
7. Bail, Christopher A; Lisa P. Argyle; Taylor W. Brown; John P. Bumpus, et al. "Exposure to Opposing Views on Social Media can Increase Political Polarization." *Proceedings of the National Academy of Sciences (PNAS)* 115, no. 37 (2018): 9216–9221. https://doi.org/10.1073/pnas.1804840115 (accessed June 17, 2026).
8. Vosoughi, Soroush; Deb Roy; and Sinan Aral. "The spread of true and false news online." *Science* 359, no. 6380 (2018): 1146–1151. https://www.science.org/doi/10.1126/science.aap9559 (accessed March 14, 2026).
9. Kaplan, Jonas T.; Sarah I. Gimbel; and Sam Harris. "Neural correlates of maintaining one's political beliefs in the face of counterevidence." *Scientific Reports* 6 (2016): 39589. https://www.nature.com/articles/srep39589 (accessed March 14, 2026).
10. Cascio, Christopher N.; Matthew Brook O'Donnell; Francis J. Tinney, Matthew D. Lieberman, et al. "Self-Affirmation Activates Brain Systems Associated with Self-Related Processing and Reward and Is Reinforced by Future Orientation," *Social Cognitive and Affective Neuroscience* 11, no. 4 (2016): 621–629. https://doi.org/10.1093/scan/nsv136 (accessed June 17, 2026).
11. Edelman. "2025 Edelman Trust Barometer: Global Report" (2025). https://www.edelman.com/sites/g/files/aatuss191/files/2025-01/2025%20Edelman%20Trust%20Barometer%20Global%20Report_01.23.25.pdf (accessed March 14, 2026).

12. Norris, Pippa. *Democratic Deficit: Critical Citizens Revisited* (Cambridge: Cambridge University Press, 2011). https://www.cambridge.org/core/books/democratic-deficit/C1A2A5421BBD8F968992706194074O5A (accessed March 14, 2026).
13. Fukuyama, Francis. *Political Order and Political Decay: From the Industrial Revolution to the Globalization of Democracy* (New York: Farrar, Straus and Giroux, 2014).
14. Duckett, Jane; and Neil Munro. 2022. "Authoritarian Regime Legitimacy and Health Care Provision: Survey Evidence from Contemporary China." *Journal of Health Politics, Policy and Law* 47, no. 3: 375–409. https://pubmed.ncbi.nlm.nih.gov/34847220/ (accessed March 14, 2026).
15. Cyberspace Administration of China et al. *Measures for the Management of Generative Artificial Intelligence Services (Draft for Comment, April 2023)*, translated and published by the Stanford University DigiChina project. https://digichina.stanford.edu/work/translation-measures-for-the-management-of-generative-artificial-intelligence-services-draft-for-comment-april-2023/ (accessed March 14, 2026).
16. Chan, Kyle; Gregory Smith; Jimmy Goodrich; Gerard DiPippo; and Konstantin F. Pilz. "Full Stack: China's Evolving Industrial Policy for AI." RAND Corporation, June 26, 2025. https://www.rand.org/pubs/perspectives/PEA4012-1.html (accessed April 13, 2026).
17. Maslej, Nestor, et al. "The AI Index 2024 Annual Report." *AI Index Steering Committee, Institute for Human-Centered AI, Stanford University*, April 2024. https://hai.stanford.edu/assets/files/hai_ai-index-report-2024-smaller2.pdf (accessed March 14, 2026); Lewis, James A. "AI Regulation Is Coming – What Is the Likely Outcome?" *Center for Strategic and International Studies (CSIS)*, October 2023. https://www.csis.org/blogs/strategic-technologies-blog/ai-regulation-coming-what-likely-outcome (accessed March 14, 2026).
18. Simonite, Tom. "How Google Plans to Solve Artificial Intelligence," *MIT Technology Review*, March 31, 2016, https://www.technologyreview.com/2016/03/31/161234/how-google-plans-to-solve-artificial-intelligence/ (accessed March 14, 2026).

19. Vaswani, Ashish; Noam Shazeer; Niki Parmar; Jakob Uszkoreit; et al. "Attention Is All You Need." *Advances in Neural Information Processing Systems* 30 (NeurIPS), 2017. https://proceedings.neurips.cc/paper_files/paper/2017/file/3f5ee243547dee91fbd053c1c4a845aa-Paper.pdf (accessed March 14, 2026).
20. *Reuters*, "New York lawyers sanctioned for using fake ChatGPT cases in legal brief," June 26, 2023. https://www.reuters.com/legal/new-york-lawyers-sanctioned-using-fake-chatgpt-cases-legal-brief-2023-06-22/ (accessed March 14, 2026).
21. McLeod, B. (2024). "B.C. tribunal confirms companies remain liable for information provided by AI chatbot." *American Bar Association – Business Law Today*. https://www.americanbar.org/groups/business_law/resources/business-law-today/2024-february/bc-tribunal-confirms-companies-remain-liable-information-provided-ai-chatbot/ (accessed March 14, 2026).
22. Feathers, Todd. "NYC's AI Chatbot Tells Businesses to Break the Law." *The Markup*, March 29, 2024. https://themarkup.org/news/2024/03/29/nycs-ai-chatbot-tells-businesses-to-break-the-law (accessed March 14, 2026).
23. Li, Y.; Y. Huang; M. Ildiz; A. Singh Rawat; and S. Oymak. *Mechanics of Next Token Prediction with Self-Attention. Proceedings of the 27th International Conference on Artificial Intelligence and Statistics (AISTATS)*, PMLR, 2024. https://proceedings.mlr.press/v238/li24f.html (accessed March 14, 2026).
24. McKinney, Scott Mayer; Marcin Sieniek; Varun Godbole; Jonathan Godwin; et al. "International Evaluation of an AI System for Breast Cancer Screening," *Nature* 577 (2020): 89–94. https://www.nature.com/articles/s41586-019-1799-6 (accessed March 14, 2026).
25. Jumper, John; Richard Evans; Alexander Pritzel; Tim Green; et al. "Highly Accurate Protein Structure Prediction with AlphaFold," *Nature* 596 (2021): 583–589. https://www.nature.com/articles/s41586-021-03819-2 (accessed March 14, 2026).
26. Chen, Mark; Jerry Tworek; Heewoo Jun; Qiming Yuan; et al. "Evaluating Large Language Models Trained on Code," *arXiv* (2021). https://arxiv.org/abs/2107.03374 (accessed March 14, 2026).

27. Wu, Yonghui; Mike Schuster; Zhifeng Chen; Quoc V. Le; et al. "Google's Neural Machine Translation System," *arXiv* (2016). https://arxiv.org/abs/1609.08144 (accessed March 14, 2026).
28. Silver, David; Aja Huang; Chris J. Maddison; Arthur Guez; et al. "Mastering the Game of Go with Deep Neural Networks and Tree Search," *Nature* 529 (2016): 484–489. https://www.nature.com/articles/nature16961 (accessed March 14, 2026).
29. Briggs, Joseph; and Devesh Kodnani. "The Potentially Large Effects of Artificial Intelligence on Economic Growth." Goldman Sachs Economics Research, March 26, 2023. https://www.gspublishing.com/content/research/en/reports/2023/03/27/d64e052b-0f6e-45d7-967b-d7be35fabd16.html (accessed March 14, 2026).
30. Challenger, Gray & Christmas. "Challenger Report: March Cuts Rise 25% From February, AI Leads Reasons." April 2, 2026. https://www.challengergray.com/blog/challenger-report-march-cuts-rise-25-from-february-ai-leads-reasons (accessed April 15, 2026).
31. Dell'Acqua, Fabrizio; Edward McFowland III; Ethan Mollick; Hila Lifshitz-Assaf, et al. "Navigating the Jagged Technological Frontier: Field Experimental Evidence of the Effects of AI on Knowledge Worker Productivity and Quality." *Harvard Business School Working Paper*, No. 24-013, 2023. https://www.hbs.edu/faculty/Pages/item.aspx?num=64700 (accessed March 14, 2026).
32. Dahmani, Louisa; and Véronique D. Bohbot. "Habitual use of GPS negatively impacts spatial memory during self-guided navigation." *Scientific Reports* 10, no. 1 (2020): 6342. https://www.nature.com/articles/s41598-020-62877-0 (accessed March 14, 2026).
33. Liu, Grace; Brian Christian; Tsvetomira Dumbalska; Michiel A. Bakker; and Rachit Dubey. "AI Assistance Reduces Persistence and Hurts Independent Performance." arXiv:2604.04721, April 6, 2026. https://arxiv.org/abs/2604.04721 (accessed June 15, 2026).
34. Rudin, Cynthia. "Stop Explaining Black Box Machine Learning Models for High Stakes Decisions and Use Interpretable Models Instead." *Nature Machine Intelligence* 1, no. 5 (2019): 206–215. https://www.nature.com/articles/s42256-019-0048-x (accessed March 14, 2026).
35. Russell, Stuart. *Human Compatible: Artificial Intelligence and the Problem of Control* (New York: Viking, 2019).

36. Regehr, K; C Shaughnessy; M Zhao; I Cambazoglu, et al. Normalizing Toxicity: The Role of Recommender Algorithms for Young People's Mental Health and Social Wellbeing. *Frontiers in Psychology* 16 (2025): 1523649. https://doi.org/10.3389/fpsyg.2025.1523649 (accessed June 17, 2026).
37. Sharma, Mrinank; Meg Tong; Tomasz Korbak; David Duvenaud; et al. "Towards Understanding Sycophancy in Language Models," Stanford University, 2023, https://arxiv.org/abs/2310.13548 (accessed March 14, 2026).
38. OpenEvidence. "OpenEvidence Announces $210 Million Round at $3.5 Billion Valuation." Press release, July 15, 2025. https://www.openevidence.com/announcements/openevidence-the-fastest-growing-application-for-physicians-in-history-announces-dollar210-million-round-at-dollar35-billion-valuation (accessed April 13, 2026).
39. Harvey. "Harvey Raises at $11 Billion Valuation to Scale Agents Across Law Firms and Enterprises." Press release, March 25, 2026. https://www.harvey.ai/blog/harvey-raises-at-dollar11-billion-valuation-to-scale-agents-across-law-firms-and-enterprises (accessed April 13, 2026).
40. Kim, Yubin; Hyewon Jeong; Shan Chen; Shuyue Stella Li; et al. Hallucination in Foundation Models and Their Challenges in Clinical Settings. medRxiv. 2025. https://www.medrxiv.org/content/10.1101/2025.02.28.25323115v2.full-text (accessed March 14, 2026).
41. Supernormal stimulus. (n.d.). *Wikipedia*. https://en.wikipedia.org/wiki/Supernormal_stimulus (accessed March 14, 2026).
42. Sharma et al. "Towards Understanding Sycophancy in Language Models." https://arxiv.org/abs/2310.13548 (accessed March 14, 2026).
43. Chen, X.; Y. Chen; L. Huang. (2024). Attachment anxiety and emotional attachment to AI companions: The mediating role of emotional bonding in problematic AI use. *Current Psychology*. https://pmc.ncbi.nlm.nih.gov/articles/PMC12379994/ (accessed March 14, 2026).
44. Reuters. "ChatGPT Sets Record for Fastest-growing User Base - Analyst Note," 2023. https://www.reuters.com/technology/chatgpt-sets-record-fastest-growing-user-base-analyst-note-2023-02-01/ (accessed June 17, 2026).
45. Quartz. "Big Tech's AI Spending Spree: $650 Billion in 1 Year," 2026. https://qz.com/big-tech-ai-capex-amazon-microsoft-meta-google (accessed June 16, 2026).

Part 2

1. Intrinsic. "Foxconn and Intrinsic Launch Joint Venture at Hon Hai Tech Day 2025." Intrinsic Blog, November 20, 2025. https://www.intrinsic.ai/blog/posts/foxconn-and-intrinsic-launch-joint-venture-to-build-the-ai-factory-of-the-future (accessed April 15, 2026).
2. United Nations Economic Commission for Africa. "As Africa's Population Crosses 1.5 Billion, The Demographic Window Is Opening." July 2024. https://www.uneca.org/stories/(blog)-as-africa%E2%80%99s-population-crosses-1.5-billion,-the-demographic-window-is-opening-getting (accessed March 14, 2026).
3. UBS Global Wealth Report 2025, as reported by Visual Capitalist. "The Global Distribution of Wealth, Shown in One Pyramid." December 2025. https://www.visualcapitalist.com/the-global-distribution-of-wealth-shown-in-one-pyramid/ (accessed March 14, 2026).
4. Fortune. "AI Talent Wars: Startups, Google, Meta, OpenAI Are Fighting for Scientists with Sky-High Pay and Stock," *Fortune*, March 15, 2025. https://fortune.com/2025/03/15/ai-talent-wars-startups-google-meta-openai-hiring-scientists-stock-salaries/ (accessed March 14, 2026); TMTPOST. "China Narrows AI Talent Gap With U.S. as Research Enters Engineering Phase: Report," https://en.tmtpost.com/post/7616647 (accessed March 14, 2026).

Chapter 4

1. The Nobel Prize. "Press release: The Nobel Prize in Physics 2024," October 8, 2024. https://www.nobelprize.org/prizes/physics/2024/press-release/ (accessed March 14, 2026).
2. The Nobel Prize. "Press release," 2024. https://www.nobelprize.org/prizes/chemistry/2024/press-release/ (accessed March 14, 2026).
3. Google DeepMind. "AlphaFold: Five years of impact," *Google DeepMind Blog*, November 25, 2025. https://deepmind.google/blog/alphafold-five-years-of-impact/ (accessed March 14, 2026).
4. Varadi, Mihály; Damian Bertoni; Paulyna Magana; Urmila Paramval, et al. "AlphaFold Protein Structure Database in 2024: Providing Structure

Coverage for Over 214 Million Protein Sequences." *Nucleic Acids Research* 52, no. D1 (2024): D368–D375. https://doi.org/10.1093/nar/gkad1011 (accessed June 17, 2026).

5. Google DeepMind. "AlphaFold 3 predicts the structure and interactions of all of life's molecules," May 8, 2024. https://blog.google/technology/ai/google-deepmind-isomorphic-alphafold-3-ai-model/ (accessed March 14, 2026).
6. Handoko, Albertus Denny; and Riko I. Made. "Artificial Intelligence and Generative Models for Materials Discovery—A Review." *arXiv*, 2025. arXiv:2508.03278. https://arxiv.org/abs/2508.03278 (accessed March 14, 2026).
7. World Economic Forum. "7 ways AI is transforming healthcare," 2025. https://www.weforum.org/stories/2025/08/ai-transforming-global-health/ (accessed March 14, 2026).
8. Microsoft AI. "The Path to Medical Superintelligence," 2024. https://microsoft.ai/news/the-path-to-medical-superintelligence/ (accessed March 14, 2026).
9. Peyrard, Maxime; and Kyunghyun Cho. "Meta-Statistical Learning: Supervised Learning of Statistical Estimators," *arXiv* 2025, https://arxiv.org/abs/2502.12088 (accessed March 16, 2026).
10. Crunchbase. "6 Charts That Show the Big AI Funding Trends of 2025," December 15, 2025. https://news.crunchbase.com/ai/big-funding-trends-charts-eoy-2025/ (accessed March 14, 2026).
11. Hayden, Jordan. "OpenAI Closes Funding Round at an $852 Billion Valuation." CNBC, March 31, 2026. https://www.cnbc.com/2026/03/31/openai-funding-round-ipo.html (accessed April 15, 2026); Novet, Jordan. "Anthropic Closes $30 Billion Funding Round at $380 Billion Valuation." CNBC, February 12, 2026. https://www.cnbc.com/2026/02/12/anthropic-closes-30-billion-funding-round-at-380-billion-valuation.html (accessed April 15, 2026).
12. Quartz. "Big Tech's AI spending spree: $650 billion in 1 year," February 2026. https://qz.com/big-tech-ai-capex-amazon-microsoft-meta-google (accessed March 14, 2026); J.P. Morgan Asset Management. "Is AI Already Driving U.S. Growth?" *J.P. Morgan Asset Management Insights*.

https://am.jpmorgan.com/us/en/asset-management/adv/insights/market-insights/market-updates/on-the-minds-of-investors/is-ai-already-driving-us-growth/ (accessed March 14, 2026).

13. Fortune. "Meta ramps up AI efforts by paying $14.3 billion for 49% of Scale AI," June 13, 2025. https://fortune.com/2025/06/13/meta-scale-ai-alexandr-wang-superintelligence-team/ (accessed March 14, 2026).
14. Meta Platforms, Inc. "Alexandr Wang, Chief AI Officer," *Meta Media Gallery*, accessed February 13, 2026, https://www.meta.com/media-gallery/executives/alexandr-wang/ (accessed March 14, 2026); CNBC, "Scale AI cuts 14% of workforce after Meta investment, hiring of founder Wang," July 16, 2025. https://www.cnbc.com/2025/07/16/scale-ai-cuts-14percent-of-workforce-after-meta-investment-hiring-of-wang.html (accessed March 14, 2026).
15. "Meta Offers $200M to Poach Apple AI Exec." *Yahoo Finance*. July 10, 2025. https://finance.yahoo.com/news/meta-offers-200m-poach-apple-104055787.html (accessed March 14, 2026).
16. Metz, Cade; and Mike Isaac. "In Pursuit of Godlike Technology, Mark Zuckerberg Amps Up the AI Race," *New York Times*, June 27, 2025. https://www.nytimes.com/2025/06/27/technology/mark-zuckerberg-meta-ai.html (accessed March 14, 2026).
17. Gartner. "Gartner Says Worldwide AI Spending Will Total $1.5 Trillion in 2025," September 17, 2025. https://www.gartner.com/en/newsroom/press-releases/2025-09-17-gartner-says-worldwide-ai-spending-will-total-1-point-5-trillion-in-2025 (accessed March 14, 2026).
18. Goldman Sachs Research. *Generative AI Could Raise Global GDP by 7%*. April 2023. https://www.goldmansachs.com/intelligence/pages/generative-ai-could-raise-global-gdp-by-7-percent.html (accessed March 14, 2026).
19. Abecasis, Manuel. "What Is the US Economy's Potential Growth Rate?" (Goldman Sachs Research, October 2025). https://www.goldmansachs.com/insights/articles/what-is-the-us-economys-potential-growth-rate (accessed March 14, 2026).
20. CNBC. "OpenAI Sees Roughly $5 Billion Loss This Year on $3.7 Billion in Revenue." September 27, 2024. https://www.cnbc.com/2024/09/27/openai-sees-5-billion-loss-this-year-on-3point7-billion-in-revenue.html (accessed March 14, 2026).

21. Smith, Dave. "OpenAI Says It Plans to Report Stunning Annual Losses Through 2028—and Then Turn Wildly Profitable Just Two Years Later." Fortune, November 12, 2025. https://fortune.com/2025/11/12/openai-cash-burn-rate-annual-losses-2028-profitable-2030-financial-documents/ (accessed April 15, 2026).
22. Nair, Devika; and Mrinmay Dey. "Anthropic Sees Revenue Potentially Soaring to $34.5 Billion in 2027, *The Information* Reports." *Reuters*, February 13, 2025; Reuters. "Anthropic Hits $3 Billion in Annualized Revenue on Business Demand for AI." May 30, 2025. https://www.reuters.com/business/anthropic-hits-3-billion-annualized-revenue-business-demand-ai-2025-05-30 (accessed April 15, 2026) https://www.reuters.com/technology/anthropic-projects-soaring-growth-345-billion-2027-revenue-information-reports-2025-02-13/ (accessed March 14, 2026).
23. Jin, Berber; and Nate Rattner. "An Inside Look at OpenAI and Anthropic's Finances Ahead of Their IPOs," *The Wall Street Journal*, April 6, 2026. https://www.wsj.com/tech/ai/openai-anthropic-ipo-finances-04b3cfb9 (accessed April 7, 2026).
24. OpenAI. "Sora 2 System Card," September 30, 2025. https://openai.com/index/sora-2-system-card/ (accessed March 14, 2026).
25. "OpenAI names ex-Facebook VP as new CEO of Applications," *Capacity*, May 19, 2025. https://capacityglobal.com/news/openai-ceo-applications/ (accessed March 14, 2026); "Fidji Simo now in charge of Facebook's app helped build a $55 billion ad business," Digiday, https://digiday.com/media/fidji-simo-now-in-charge-of-facebooks-app-helped-facebook-build-a-55-billion-ad-business/ (accessed April 15, 2026).
26. "OpenAI Searching for Ad Chief to Bolster Monetization." *PYMNTS*. September 24, 2025. https://www.pymnts.com/artificial-intelligence-2/2025/openai-searching-for-ad-chief-to-bolster-monetization/ (accessed March 14, 2026).
27. Jin, Berber; and Jessica Toonkel. "The Sudden Fall of OpenAI's Most Hyped Product Since ChatGPT." The Wall Street Journal, March 30, 2026. https://www.wsj.com/tech/ai/the-sudden-fall-of-openais-most-hyped-product-since-chatgpt-64c730c9 (accessed April 15, 2026).

28. Gartner. "The 2025 Hype Cycle for Artificial Intelligence Goes Beyond GenAI," August 2025. https://www.gartner.com/en/articles/hype-cycle-for-artificial-intelligence (accessed March 14, 2026).
29. S&P Global Market Intelligence. "AI Experiences Rapid Adoption but with Mixed Outcomes—Highlights from Voice of the Enterprise: AI & Machine Learning," May 30, 2025. https://www.spglobal.com/market-intelligence/en/news-insights/research/ai-experiences-rapid-adoption-but-with-mixed-outcomes-highlights-from-vote-ai-machine-learning (accessed March 14, 2026).
30. Writer | Workplace Intelligence. *Generative AI Adoption in the Enterprise: 2025 Survey*. December 2024. https://go.writer.com/hubfs/pdfs/generative-ai-adoption-enterprise-survey-writer-com.pdf (accessed March 14, 2026).
31. World Economic Forum. "The Future of Jobs Report 2025," January 2025. https://reports.weforum.org/docs/WEF_Future_of_Jobs_Report_2025.pdf (accessed March 14, 2026).
32. Tech Policy Press. "Transcript: Sam Altman Testifies at US Senate Hearing on AI Competitiveness." May 8, 2025. https://www.techpolicy.press/transcript-sam-altman-testifies-at-us-senate-hearing-on-ai-competitiveness (accessed April 15, 2026).
33. The White House. "Removing Barriers to American Leadership in Artificial Intelligence," January 23, 2025. https://www.whitehouse.gov/presidential-actions/2025/01/removing-barriers-to-american-leadership-in-artificial-intelligence/ (accessed March 14, 2026).
34. Soper, Spencer; and Josh Wingrove. "Head of U.S. AI Safety Institute to Leave as Trump Shifts Course," *Bloomberg*, February 4, 2025. https://www.bloomberg.com/news/articles/2025-02-04/head-of-us-ai-safety-institute-to-leave-as-trump-shifts-course (accessed March 14, 2026).
35. Allen, Gregory C. "Choking Off China's Access to the Future of AI," *Center for Strategic and International Studies*, October 11, 2022. https://www.csis.org/analysis/choking-chinas-access-future-ai (accessed March 14, 2026).
36. "DeepSeek: A Deep Dive." *Center for Strategic and International Studies* (CSIS), April 8, 2025. https://www.csis.org/analysis/deepseek-deep-dive (accessed March 14, 2026).
37. "Nvidia Sheds Record $589 Billion in Value as China's DeepSeek Sparks AI Rout," *Bloomberg*, January 27, 2025. https://www.bloomberg.com/

news/articles/2025-01-27/asml-sinks-as-china-ai-startup-triggers-panic-in-tech-stocks (accessed March 14, 2026).

38. "The Global AI Talent Tracker 2.0." MacroPolo, Paulson Institute, 2023. https://macropolo.org/digital-projects/the-global-ai-talent-tracker/ (accessed March 14, 2026).
39. Kerr, Dara. "Trump clears way for Nvidia to sell powerful AI chips to China," *The Guardian*, December 9, 2025. https://www.theguardian.com/technology/2025/dec/08/trump-nvidia-ai-chips-china (accessed March 14, 2026).
40. "U.S. Authorities Shut Down Major China-Linked AI Tech Smuggling Network," U.S. Department of Justice, December 8, 2025. https://www.justice.gov/opa/pr/us-authorities-shut-down-major-china-linked-ai-tech-smuggling-network (accessed March 14, 2026).
41. *Interim Measures for the Management of Generative AI Services. Wikipedia*. https://en.wikipedia.org/wiki/Interim_Measures_for_the_Management_of_Generative_AI_Services (accessed March 14, 2026).
42. European Commission. "EU Artificial Intelligence Act." https://digital-strategy.ec.europa.eu/en/policies/regulatory-framework-ai (accessed March 14, 2026).
43. "Artificial Intelligence Action Summit," Élysée Palace, February 11, 2025. https://www.elysee.fr/admin/upload/default/0001/17/786758b38da7b4c16f26dc56e51884b3346684aa.pdf (accessed March 14, 2026); "Macron Touts $109 Billion of Investments in Bid to Become AI Hub," *Bloomberg*, February 10, 2025. https://www.bloomberg.com/news/articles/2025-02-10/macron-touts-109-billion-of-investments-in-bid-to-become-ai-hub (accessed March 14, 2026).
44. "SG Announces New AI Safety Initiatives at the Global AI Action Summit in France," *Infocomm Media Development Authority (IMDA)*, February 11, 2025. https://www.imda.gov.sg/resources/press-releases-factsheets-and-speeches/press-releases/2025/singapore-ai-safety-initiatives-global-ai-summit-france (accessed March 14, 2026).
45. Personal Data Protection Commission (PDPC) and Infocomm Media Development Authority (IMDA). *Model AI Governance Framework 2.0*. Singapore: PDPC/IMDA, January 2020. https://www.pdpc.gov.sg/-/media/files/pdpc/pdf-files/resource-for-organisation/ai/sgmodelaigovframework2.pdf (accessed March 14, 2026).

46. AI Verify Foundation. "What Is AI Verify?" https://aiverifyfoundation.sg/what-is-ai-verify/ (accessed March 14, 2026).
47. Russell, Stuart. *Human Compatible: Artificial Intelligence and the Problem of Control* (New York: Viking, 2019).
48. Russell, Stuart. "Provably Beneficial Artificial Intelligence," in *The Next Step: Exponential Life*, BBVA OpenMind, 2017. https://people.eecs.berkeley.edu/~russell/papers/russell-bbvabook17-pbai.pdf (accessed March 14, 2026).
49. "Three Mile Island Accident." *Wikipedia*. https://en.wikipedia.org/wiki/Three_Mile_Island_accident (accessed March 14, 2026).
50. "International Dialogues on AI Safety." *Safe AI Forum*, October 2023. https://idais.ai/ (accessed March 14, 2026).
51. China Daily. "Special Attention Needed to Ensure AI Safety, US Professor Says," January 14, 2025. https://www.chinadaily.com.cn/a/202501/14/WS678643c2a310f1265a1dacb9.html (accessed March 14, 2026).
52. Vaswani, Ashish; Noam Shazeer; Niki Parmar, et al. "Attention Is All You Need," *Advances in Neural Information Processing Systems (NeurIPS)*, 30 (2017). https://arxiv.org/abs/1706.03762 (accessed March 14, 2026).

Chapter 5

1. "Inside Spotify's Recommendation System: A Complete Guide." *Music Tomorrow*, last modified 2025. https://www.music-tomorrow.com/blog/how-spotify-recommendation-system-works-complete-guide (accessed March 14, 2026).
2. Jeyachandran, Satish. "Introducing the 5th-Generation Waymo Driver," *Waymo Blog*, March 4, 2020, https://waymo.com/blog/2020/03/introducing-5th-generation-waymo-driver (accessed March 14, 2026).
3. "Waymo Safety Impact." *Waymo*. https://waymo.com/safety/impact/ (accessed March 14, 2026).
4. Talaviya, Tanha; Dhara Shah; Nivedita Patel; Hiteshri Yagnik; et al. "Implementation of Artificial Intelligence in Agriculture for Optimisation of Irrigation and Application of Pesticides and Herbicides." *Artificial Intelligence in Agriculture*, 4 (2020): 58–73. https://doi.org/10.1016/j.aiia.2020.04.002 (accessed June 17, 2026); Izquierdo-Bueno, Inmaculada,

Moraga Javier, Cantoral Jesús M, Carbú María, et al. "Smart Viniculture: Applying Artificial Intelligence for Improved Winemaking and Risk Management." *Applied Sciences* 14, no. 22 (2024): 10277. https://doi.org/10.3390/app142210277 (accessed June 17, 2026).

5. Barbedo, Jayme Garcia Arnal. "A Review on the Use of Unmanned Aerial Vehicles and Imaging Sensors for Monitoring and Assessing Plant Stresses." *Drones* 3, no. 2 (2019): 40. https://www.mdpi.com/2504-446X/3/2/40 (accessed March 14, 2026).
6. Fountas, Spyros; Nikos Mylonas; Ioannis Malounas; Efthymios Rodias; et al. "Agricultural Robotics for Field Operations." *Sensors* 20, no. 9 (2020): 2672. https://www.mdpi.com/1424-8220/20/9/2672 (accessed March 14, 2026).
7. Elgalb, Ahmed; and Maher Gerges. "Optimizing Supply Chain Logistics with Big Data and AI: Applications for Reducing Food Waste." *Journal of Current Science and Research Review* 2, no. 02 (2024): 29–39. https://jcsrr.org/index.php/jcsrr/article/view/68 (accessed March 14, 2026).
8. "Mastercard Supercharges Consumer Protection with Gen AI." *Mastercard Newsroom*, February 1, 2024, https://www.mastercard.com/global/en/news-and-trends/press/2024/february/mastercard-supercharges-consumer-protection-with-gen-ai.html (accessed March 14, 2026).
9. Zoldi, Scott. "Evolution of Fraud Analytics – An Inside Story," *KDnuggets*, March 2014. https://www.kdnuggets.com/2014/03/evolution-fraud-analytics-inside-story.html (accessed March 14, 2026).
10. "The History of the FICO® Score." *myFICO Blog*, August 21, 2018. https://www.myfico.com/credit-education/blog/history-of-the-fico-score (accessed March 14, 2026).
11. "Energy and AI." *International Energy Agency (IEA)*, April 2025, https://www.iea.org/reports/energy-and-ai (accessed March 14, 2026); https://iea.blob.core.windows.net/assets/de9dea13-b07d-42c5-a398-d1b3ae17d866/EnergyandAI.pdf (accessed March 14, 2026).
12. "High-Frequency Trading." *Wikipedia*. https://en.wikipedia.org/wiki/High-frequency_trading (accessed March 14, 2026).
13. Altman, Sam, quoted in TechCrunch. "Sam Altman says ChatGPT has hit 800M weekly active users," October 6, 2025. https://techcrunch.

com/2025/10/06/sam-altman-says-chatgpt-has-hit-800m-weekly-active-users/ (accessed March 14, 2026).

14. Elon University. "Survey: 52% of U.S. adults now use AI large language models like ChatGPT," March 12, 2025. https://imaginingthedigitalfuture.org/wp-content/uploads/2025/03/ITDF-LLM-User-Report-3-12-25.pdf (accessed March 14, 2026).
15. OpenAI. "Sora 2 is here," September 30, 2025. https://openai.com/index/sora-2/ (accessed March 14, 2026).
16. VentureBeat. "OpenAI debuts Sora 2 AI video generator app with sound and self-insertion cameos, API coming soon," October 1, 2025. https://venturebeat.com/ai/openai-debuts-sora-2-ai-video-generator-app-with-sound-and-self-insertion (accessed March 14, 2026).
17. CNBC. "OpenAI's Sora 2 is putting safety and censorship to the test with stunningly real videos," October 2, 2025, https://www.cnbc.com/2025/10/02/openai-invite-code-sora-2-censorship.html (accessed March 14, 2026); Christopher, Nilesh, "Sora App's Hyperreal AI Videos Ignite Online Trust Crisis as Downloads Surge," *Los Angeles Times*, October 26, 2025. https://www.latimes.com/business/story/2025-10-26/sora-the-bizarre-mind-bending-ai-slop-machine (accessed March 14, 2026).
18. Tech Times. "Meta's new 'vibes' feed turns TikTok into AI slop, But who actually wants it?" September 26, 2025. https://www.techtimes.com/articles/312083/20250926/metas-new-vibes-feed-turns-tiktok-ai-slop-who-actually-wants-it.htm (accessed March 14, 2026).
19. Jin, Berber; and Jessica Toonkel. "The Sudden Fall of OpenAI's Most Hyped Product Since ChatGPT." *The Wall Street Journal*, March 30, 2026. https://www.wsj.com/tech/ai/the-sudden-fall-of-openais-most-hyped-product-since-chatgpt-64c730c9 (accessed April 15, 2026).
20. Fortune. "Satya Nadella on the 'enigma of success' in the age of AI: A thriving business, but 15,000+ layoffs," July 24, 2025. https://fortune.com/2025/07/24/satya-nadella-memo-microsoft-layoffs-enigma-of-success-artificial-intelligence/ (accessed March 14, 2026).
21. CNBC. "AI will shrink Amazon's workforce in the coming years, CEO Jassy says," June 17, 2025. https://www.cnbc.com/2025/06/17/ai-amazon-workforce-jassy.html (accessed March 14, 2026).

22. Fortune. "Mark Zuckerberg warned employees to prepare for an 'intense year.' He just laid off 3,600 workers—including some who claim they got glowing reviews," February 11, 2025. https://fortune.com/2025/02/11/mark-zuckerberg-meta-layoffs-low-performers/ (accessed March 14, 2026); Quicker Better Tech (Forbes contributor). "Zuckerberg Says AI Will Replace Mid-Level Engineers Soon." *Forbes*, January 26, 2025. https://www.forbes.com/sites/quickerbettertech/2025/01/26/business-tech-news-zuckerberg-says-ai-will-replace-mid-level-engineers-soon/ (accessed March 14, 2026).
23. Xinhua / China National Energy Administration. "Renewables Account for Over 60 Pct of China's Power Capacity in 2025." January 30, 2026. https://english.www.gov.cn/archive/statistics/202601/30/content_WS697cb463c6d00ca5f9a08da7.html (accessed March 14, 2026).
24. SolarPower Europe. "New Report: EU Hits 2025 Solar Target but Market Contraction Puts 2030 Goal at Risk." December 2025. https://www.solarpowereurope.org/press-releases/new-report-eu-hits-2025-solar-target-but-market-contraction-puts-2030-goal-at-risk (accessed March 14, 2026).
25. Solar Energy Industries Association. "U.S. Adds 11.7 GW of New Solar Capacity in Q3, Third Largest Quarter on Record." December 9, 2025. https://seia.org/news/third-largest-quarter-on-record/ (accessed March 14, 2026); DiGangi, Diana. "2025 on Track for Record Storage Deployments: WoodMac." *Utility Dive*, December 17, 2025. https://www.utilitydive.com/news/energy-storage-installations-capacity-battery-obbba/808115/ (accessed March 14, 2026).
26. Mastercard Newsroom. "Mastercard Supercharges Consumer Protection with Gen AI." https://www.mastercard.com/news/press/2024/february/mastercard-supercharges-consumer-protection-with-gen-ai/ (accessed March 14, 2026).
27. "Mastercard." World Food Programme. https://www.wfp.org/partners/mastercard (accessed March 14, 2026).
28. Mastercard. "Mastercard Acquires Brighterion," Press Release, 2017. https://investor.mastercard.com/investor-news/investor-news-details/2017/Mastercard-Enhances-Artificial-Intelligence-Capability-with-the-Acquisition-of-Brighterion-Inc/default.aspx (accessed March 14, 2026).
29. Retail Dive. "Mastercard Acquires Behavioral Biometrics Security Firm NuData," April 2017. https://investor.mastercard.com/investor-news/

investor-news-details/2017/Mastercard-Enhances-Security-of-the-Internet-of-Things-with-the-Acquisition-of-NuData-Security-Inc/default.aspx (accessed March 14, 2026).

30. "Mastercard to Acquire Finicity to Advance Open Banking Strategy." *Mastercard Newsroom*, June 23, 2020. https://www.mastercard.com/us/en/news-and-trends/Insights/2020/Mastercard-to-Acquire-Finicity-to-Advance-Open-Banking-Strategy.html (accessed March 14, 2026).
31. "Mastercard completes acquisition of Dynamic Yield." *Dynamic Yield Blog*, April 1, 2022. https://www.dynamicyield.com/blog/mastercard-completes-acquisition-of-dynamic-yield/ (accessed March 14, 2026).
32. Mastercard Newsroom. "Mastercard Supercharges Consumer Protection with Gen AI."
33. Isik, Öykü; and Lisa Shea. "Mastercard's Ethical Approach to Governing AI," IMD Business School Case Study IM1225, 2022. https://www.imd.org/ibyimd/innovation/lessons-from-a-pioneer-mastercards-experience-of-ai-governance (accessed April 15, 2026).
34. Mastercard Press Release. "Mastercard unveils Agent Pay, pioneering agentic payments technology to power commerce in the age of AI," April 29, 2025. https://www.mastercard.com/global/en/news-and-trends/press/2025/april/mastercard-unveils-agent-pay-pioneering-agentic-payments-technology-to-power-commerce-in-the-age-of-ai.html (accessed March 14, 2026).
35. Salesforce. "Customer Success Statistics," Salesforce official website. https://www.salesforce.com/company/our-story/ (accessed March 14, 2026).
36. Benioff, Marc; and Carlye Adler. *Behind the Cloud: The Untold Story of How Salesforce.com Went from Idea to Billion-Dollar Company-and Revolutionized an Industry* (San Francisco: Jossey-Bass, 2009).
37. Salesforce. "How the AppExchange Security Review Works." *ISVforce Guide*. https://developer.salesforce.com/docs/atlas.en-us.packagingGuide.meta/packagingGuide/security_review_how_it_works.htm (accessed March 14, 2026).
38. Salesforce. (n.d.). *Einstein Trust Layer. Salesforce Help*. https://help.salesforce.com/s/articleView?id=sf.generative_ai_trust_layer.htm (accessed March 14, 2026).

Chapter 6

1. MIT AI Risk Repository. https://airisk.mit.edu/ (accessed March 15, 2026).
2. European Commission. "EU AI Act: Risk-based Approach." https://www.euaiact.com/key-issue/3 (accessed March 15, 2026).
3. National Institute of Standards and Technology (NIST). "AI Risk Management Framework." https://www.nist.gov/itl/ai-risk-management-framework (accessed March 15, 2026).
4. Bengio, Y., et al. "International AI Safety Report," January 2025. https://internationalaisafetyreport.org/sites/default/files/2025-10/international_ai_safety_report_2025_english.pdf (accessed March 15, 2026).
5. Zeng, Yi; Kevin Klyman; Andy Zhou; Yu Yang; et al. AI Risk Categorization Decoded (AIR 2024): From Government Regulations to Corporate Policies, 2024. https://arxiv.org/abs/2406.17864 (accessed March 15, 2026).
6. Anthropic. Anthropic's Responsible Scaling Policy, 2023. https://www.anthropic.com/news/anthropics-responsible-scaling-policy (accessed March 15, 2026).
7. OpenAI. *Preparedness Framework: Version 2*. April 15, 2025. *OpenAI*. https://cdn.openai.com/pdf/18a02b5d-6b67-4cec-ab64-68cdfbddebcd/preparedness-framework-v2.pdf (accessed March 15, 2026).
8. Google DeepMind. Introducing the Frontier Safety Framework, 2024. https://deepmind.google/discover/blog/introducing-the-frontier-safety-framework/ (accessed March 15, 2026).
9. Dastin, Jeffrey. "Amazon scraps secret AI recruiting tool that showed bias against women," *Reuters*, October 2018. https://www.reuters.com/article/us-amazon-com-jobs-automation-insight-idUSKCN1MK08G/ (accessed March 15, 2026).
10. Obermeyer, Ziad; Brian Powers; Christine Vogeli; and Sendhil Mullainathan. "Dissecting racial bias in an algorithm used to manage the health of populations," *Science* 366, no. 6464 (2019): 447–453. https://www.science.org/doi/10.1126/science.aax2342 (accessed March 15, 2026).

11. Wilson, Kyra; and Aylin Caliskan. "Gender, Race, and Intersectional Bias in Resume Screening via Language Model Retrieval," *Proceedings of the AAAI/ACM Conference on AI, Ethics, and Society* 7, no. 1 (2024): 1578–1590. https://doi.org/10.1609/aies.v7i1.31748 (accessed June 17, 2026).
12. Vaswani, Ashish; Noam Shazeer; Niki Parmar; Jakob Uszkoreit, et al. "Attention Is All You Need," 2017. https://arxiv.org/abs/1706.03762 (accessed March 15, 2026).
13. Digital Forensic Research Lab (DFRLab). (2025). "Exposing Pravda: How pro-Kremlin forces are poisoning AI models and rewriting Wikipedia." *Atlantic Council.* https://www.atlanticcouncil.org/blogs/new-atlanticist/exposing-pravda-how-pro-kremlin-forces-are-poisoning-ai-models-and-rewriting-wikipedia/ (accessed March 15, 2026).
14. Feldstein, Steven. *AI & Big Data Global Surveillance Index (2022 updated), Version 4.* Mendeley Data, June 6, 2022. https://doi.org/10.17632/gjhf5y4xjp.4 (accessed June 17, 2026).
15. Cerullo, Megan. "AI Is Leading to Thousands of Job Losses, Report Finds," *CBS News*, August 5, 2025. https://www.cbsnews.com/news/ai-jobs-layoffs-us-2025/ (accessed March 15, 2026).
16. Economic Policy Institute. "How Should We Assess and Characterize Worker Wage Growth in Recent Decades?" April 2025. https://www.epi.org/blog/how-should-we-assess-and-characterize-workers-wage-growth-in-recent-decades/ (accessed March 15, 2026).
17. Economic Policy Institute. "Growing Inequalities, Reflecting Growing Employer Power, Have Generated a Productivity–Pay Gap Since 1979," August 2020. https://www.epi.org/blog/growing-inequalities-reflecting-growing-employer-power-have-generated-a-productivity-pay-gap-since-1979-productivity-has-grown-3-5-times-as-much-as-pay-for-the-typical-worker/ (accessed March 15, 2026).
18. Google. "Our Third Decade of Climate Action: Realizing a Carbon-Free Future," September 14, 2020. https://blog.google/outreach-initiatives/sustainability/our-third-decade-climate-action-realizing-carbon-free-future/ (accessed March 15, 2026); Microsoft, "2024 Environmental Sustainability Report," https://www.microsoft.com/en-us/corporate-responsibility/sustainability (accessed March 15, 2026).

19. NPR. "AI brings soaring emissions for Google and Microsoft," July 12, 2024. https://www.npr.org/2024/07/12/g-s1-9545/ai-brings-soaring-emissions-for-google-and-microsoft (accessed March 15, 2026).
20. Cloudflare. "DDoS Threat Report for 2025 Q3," December 2025. https://blog.cloudflare.com/ddos-threat-report-2025-q3/ (accessed March 15, 2026); Toulas, Bill. "The Biggest DDoS Attack Ever Has Been Detected - but Fortunately You Probably Barely Noticed It." *TechRadar*, January 29, 2026. https://www.techradar.com/pro/security/the-biggest-ddos-attack-ever-has-been-detected-but-fortunately-you-probably-barely-noticed-it (accessed March 15, 2026).
21. Cloudflare. "DDoS Threat Report for 2025 Q3."
22. Cloudflare. *Signals Report: Resilience at Scale* / theNET, "Combating AI-Powered Threats," Mike Hamilton (CIO, Cloudflare), November 2025. https://www.cloudflare.com/the-net/security-signals/ai-powered-threats/ (accessed March 15, 2026).
23. Zetter, Kim. "Hackers Detail How They Allegedly Stole Ticketmaster Data From Snowflake," *WIRED*, June 17, 2024. https://www.wired.com/story/epam-snowflake-ticketmaster-breach-shinyhunters/ (accessed March 15, 2026).
24. NHS England. "Synnovis Ransomware Cyber-Attack," June 2024. https://www.england.nhs.uk/london/synnovis-ransomware-cyber-attack/ (accessed March 15, 2026); Paganini, Pierluigi. "London hospitals canceled over 800 operations in the week after Synnovis ransomware attack," *SecurityAffairs*, June 15, 2024. https://securityaffairs.com/164541/cyber-crime/londol-hospitals-canceled-800-operations-ransomware.html (accessed March 15, 2026); American Hospital Association. "Change Healthcare Cyberattack Underscores Urgent Need to Strengthen Cyber Preparedness," 2025. https://www.aha.org/change-healthcare-cyberattack-underscores-urgent-need-strengthen-cyber-preparedness-individual-health-care-organizations-and (accessed March 15, 2026).
25. Singer, Brian; Keane Lucas; Lakshmi Adiga; Meghna Jain, et al. "On the Feasibility of Using LLMs to Autonomously Execute Multi-host Network Attacks," arXiv preprint arXiv:2501.16466, January 2025. https://arxiv.org/abs/2501.16466 (accessed March 15, 2026).

26. Cisco. "Cybersecurity Readiness Index 2025," 2025. https://newsroom.cisco.com/c/r/newsroom/en/us/a/y2025/m05/cybersecurity-readiness-index-2025.html (accessed March 15, 2026).
27. Research and Markets. *Artificial Intelligence in Cybersecurity Market*. https://www.researchandmarkets.com/report/artificial-intelligence-based-cybersecurity (accessed April 12, 2026).
28. Colman, Ben. "Why Detecting Dangerous AI Is Key to Keeping Trust Alive in the Deepfake Era," *World Economic Forum*, July 7, 2025. https://www.weforum.org/stories/2025/07/why-detecting-dangerous-ai-is-key-to-keeping-trust-alive/ (accessed March 15, 2026).
29. Gartner, Inc., "Gartner Survey Reveals GenAI Attacks Are on the Rise," press release, September 22, 2025, https://www.gartner.com/en/newsroom/press-releases/2025-09-22-gartner-survey-reveals-generative-artificial-intelligence-attacks-are-on-the-rise (accessed March 15, 2026).
30. Group-IB. "The Anatomy of a Deepfake Voice Phishing Attack: How AI-Generated Voices Are Powering the Next Wave of Scams," *Group-IB Blog*, August 6, 2025. https://www.group-ib.com/blog/voice-deepfake-scams/ (accessed March 15, 2026).
31. World Economic Forum. "Cybercrime: Lessons learned from a $25m deepfake attack," February 4, 2025. https://www.weforum.org/stories/2025/02/deepfake-ai-cybercrime-arup/ (accessed March 15, 2026); Professional Risk Managers' International Association (PRMIA). (2025). *The Arup deepfake fraud: A PRMIA case study* (pp. 1–7). PRMIA. https://prmia.org/common/Uploaded%20files/eCyber/PRMIA%20Case%20study%20-%20ARUP.pdf (accessed March 15, 2026).
32. Federal Communications Commission. "FCC Makes AI-Generated Robocalls Illegal," February 2024. https://www.fcc.gov/document/fcc-makes-ai-generated-voices-robocalls-illegal (accessed March 15, 2026).
33. IBM X-Force. "AI vs. Human Deceit: Unravelling the New Age of Phishing Tactics," October 2023; IBM, "X-Force Threat Intelligence Index 2024," February 2024. https://www.ibm.com/think/x-force/ai-vs-human-deceit-unravelling-new-age-phishing-tactics (accessed March 15, 2026).
34. KnowBe4. *Phishing Threat Trends Report 2025*. Vol. 6. Tampa, FL: KnowBe4, March 1, 2025. https://www.knowbe4.com/hubfs/Phishing-Threat-Trends-2025_Report.pdf (accessed March 15, 2026).

35. Heiding, Fred; Simon Lermen; Andrew Kao; Bruce Schneier, et al. "Evaluating Large Language Models' Capability to Launch Fully Automated Spear Phishing Campaigns: Validated on Human Subjects." arXiv preprint arXiv:2412.00586 (November 2024). https://arxiv.org/abs/2412.00586 (accessed March 15, 2026).
36. McKinsey & Company, "The Cybersecurity Provider's Next Opportunity: Making AI Safer," November 14, 2024. https://www.mckinsey.com/capabilities/risk-and-resilience/our-insights/the-cybersecurity-providers-next-opportunity-making-ai-safer (accessed April 13, 2026).
37. Anthropic. "Detecting and countering misuse of AI: August 2025." https://www.anthropic.com/news/detecting-countering-misuse-aug-2025 (accessed March 15, 2026).
38. NYU Tandon School of Engineering. "Large language models can execute complete ransomware attacks autonomously," August 2025. https://engineering.nyu.edu/news/large-language-models-can-execute-complete-ransomware-attacks-autonomously-nyu-tandon-research (accessed March 15, 2026).
39. Wack, Morgan; Carl Ehrett; Darren Linvill; Patrick Warren, et al. "Generative propaganda: Evidence of AI's impact from a state-backed disinformation campaign," *PNAS Nexus* 4, no. 4 (2025): pgaf083. https://academic.oup.com/pnasnexus/article/4/4/pgaf083/8097936 (accessed March 15, 2026).
40. Bengio et al. "International AI Safety Report."
41. Russell. *Human Compatible: Artificial Intelligence and the Problem of Control.*
42. Mata v. Avianca, Inc., No. 22-cv-1461 (S.D.N.Y. June 22, 2023). https://caselaw.findlaw.com/court/us-dis-crt-sd-new-yor/2335142.html (accessed March 15, 2026).
43. OpenAI. "Why Language Models Hallucinate," September 2025. https://openai.com/index/why-language-models-hallucinate/ (accessed March 15, 2026).
44. Zhou, Kaitlyn; Jena D. Hwang; Xiang Ren; and Maarten Sap. "Relying on the Unreliable: The Impact of Language Models' Reluctance to Express Uncertainty." In *Proceedings of the 62nd Annual Meeting of the Association for Computational Linguistics*, 3623–3643. Bangkok:

Association for Computational Linguistics, 2024. https://arxiv.org/pdf/2401.06730 (accessed March 15, 2026); Rathi, Neil; Dan Jurafsky; and Kaitlyn Zhou. "Humans Overrely on Overconfident Language Models, across Languages." Conference on Language Modeling (COLM), 2025. https://openreview.net/pdf?id=QsQatTzATT (accessed March 15, 2026).

45. Gillespie, Nicole; Steven Lockey; Tabi Ward; A. Macdade, et al. "Trust, Attitudes and Use of Artificial Intelligence: A Global Study 2025," *University of Melbourne and KPMG*, May 2025. https://assets.kpmg.com/content/dam/kpmgsites/xx/pdf/2025/05/trust-attitudes-and-use-of-ai-global-report.pdf (accessed March 15, 2026).
46. Kosmyna, Nataliya; Eugene Hauptmann; Ye Tong Yuan; Jessica Situ, et al. "Your Brain on ChatGPT: Accumulation of Cognitive Debt When Using an AI Assistant for Essay Writing Task," arXiv preprint arXiv:2506.08872 (2025). https://doi.org/10.48550/arXiv.2506.08872 (accessed June 17, 2026).
47. Sharma, Mrinank; Meg Tong; Tomasz Korbak; David Duvenaud; et al. "Towards Understanding Sycophancy in Language Models," *arXiv*, arXiv:2310.13548, 2023. https://arxiv.org/abs/2310.13548 (accessed March 15, 2026).
48. Anthropic. "Natural Emergent Misalignment from Reward Hacking," 2025. https://assets.anthropic.com/m/74342f2c96095771/original/Natural-emergent-misalignment-from-reward-hacking-paper.pdf (accessed March 15, 2026).
49. OpenAI. *GPT-4 System Card* (pp. 15–16). OpenAI, 2023. https://cdn.openai.com/papers/gpt-4-system-card.pdf (accessed March 15, 2026).
50. Greenblatt, Ryan; Carson Denison; Benjamin Wright; Fabien Roger, et al. (2024). "Alignment Faking in Large Language Models." *Anthropic & Redwood Research*. https://arxiv.org/abs/2412.14093 (accessed March 15, 2026).
51. Greenblatt et al. "Alignment Faking in Large Language Models."
52. Apollo Research. *Frontier Models are Capable of In-context Scheming*. Apollo Research, 2024. https://www.apolloresearch.ai/research/scheming-reasoning-evaluations (accessed March 15, 2026).

53. Apollo Research. *Frontier Models are Capable of In-context Scheming.*
54. Schoen, Bronson; Evgenia Nitishinskaya; Mikita Balesni; Axel Højmark, et al. "Stress Testing Deliberative Alignment for Anti-Scheming Training." arXiv preprint arXiv:2509.15541, September 19, 2025. https://arxiv.org/abs/2509.15541 (accessed March 15, 2026).
55. Mikulik, Vlad; Joe Benton; Benjamin Wright; et al. "Towards Training-Time Mitigations for Alignment Faking in RL." Anthropic Alignment Science, December 16, 2025. https://alignment.anthropic.com/2025/alignment-faking-mitigations/ (accessed March 15, 2026).
56. Gu, Albert; and Tri Dao. (2023). *Mamba: Linear-Time Sequence Modeling with Selective State Spaces.* https://arxiv.org/abs/2312.00752 (accessed March 15, 2026).
57. Garcez, Artur d'Avila; and Luís C. Lamb. Neurosymbolic AI: The 3rd Wave. *Artificial Intelligence Review* 56, no. 11 (2023): 12387–12406. https://link.springer.com/article/10.1007/s10462-023-10448-w (accessed March 15, 2026).
58. Office of the Surgeon General. (2023). Our Epidemic of Loneliness and Isolation: The U.S. Surgeon General's Advisory on the Healing Effects of Social Connection and Community. U.S. Department of Health and Human Services. https://www.hhs.gov/sites/default/files/surgeon-general-social-connection-advisory.pdf (accessed March 15, 2026).
59. Rogers, Reece. "Character.AI Gave Up on AGI. Now It's Selling Stories." *Wired*, August 12, 2025. https://www.wired.com/story/character-ai-ceo-chatbots-entertainment (accessed March 15, 2026).
60. Garcia v. Character.AI. Complaint for Damages, Case No. 6:24-cv-01903 (M.D. Fla. October 22, 2024). https://cdn.arstechnica.net/wp-content/uploads/2024/10/Garcia-v-Character-Technologies-Complaint-10-23-24.pdf (accessed March 15, 2026).
61. Raine v. OpenAI, Inc. Complaint for Damages (Cal. Super. Ct., S.F. Cty. filed August 26, 2025). https://www.courthousenews.com/wp-content/uploads/2025/08/raine-vs-openai-et-al-complaint.pdf (accessed March 15, 2026).
62. First County Bank v. OpenAI Foundation. Complaint for Damages, Case No. CGC-25-612345 (Cal. Super. Ct., S.F. Cty. filed Dec. 11, 2025). https://

www.courthousenews.com/wp-content/uploads/2025/12/ChatGPT-lawsuit-SF.pdf (accessed March 15, 2026).

63. Wikipedia. (2025). *Deaths linked to chatbots*. Wikipedia. https://en.wikipedia.org/wiki/Deaths_linked_to_chatbots (accessed March 15, 2026).
64. Social Media Victims Law Center. "Seven lawsuits filed against OpenAI," November 2025. https://socialmediavictims.org/press-releases/ (accessed March 15, 2026); Penley, Taylor. "Texas family sues Character. AI after chatbot allegedly encouraged autistic son to harm parents and himself." *Fox News*, November 18, 2025. https://www.foxnews.com/media/texas-family-sues-character-ai-after-chatbot-allegedly-encouraged-teen-harm-parents-himself (accessed March 15, 2026).
65. Garcia v. Character Technologies, Inc., Order on Motion to Dismiss, No. 6:24-cv-1903-ACC-UAM (M.D. Fla. May 21, 2025). https://scholarblogs.emory.edu/proflawrence/files/2025/05/Garcia-v.-Character-Technologies-Inc.-et-al-Entry-115.pdf (accessed March 15, 2026).
66. Fang, Cathy Mengying; Auren R. Liu; Valdemar Danry; Eunhae Lee, et al. (2025). How AI and Human Behaviors Shape Psychosocial Effects of Chatbot Use: A Longitudinal Randomized Controlled Study. arXiv. https://arxiv.org/abs/2503.17473 (accessed March 15, 2026).
67. Dohmke, T. (2023). "GitHub Copilot for Business is Now Available." *GitHub Blog*. https://github.blog/news-insights/product-news/github-copilot-for-business-is-now-available/ (accessed March 15, 2026).
68. Gillespie et al. "Trust, Attitudes and Use of Artificial Intelligence: A Global Study 2025."
69. Humlum, Anders; and Emilie Vestergaard. (2025). *Large Language Models, Small Labor Market Effects*. National Bureau of Economic Research (NBER). https://www.nber.org/papers/w33777 (accessed March 15, 2026).
70. CNBC. "OpenAI dissolves Superalignment team," May 2024. https://www.cnbc.com/2024/05/17/openai-superalignment-sutskever-leike.html (accessed March 15, 2026).
71. O'Sullivan, Isobel. (2024). *Half of OpenAI's Safety Team Quit as Concerns Over AGI Mount*. Tech.co. https://tech.co/news/half-of-openais-security-team-quit (accessed March 15, 2026).

72. O'Sullivan. *Half of OpenAI's Safety Team Quit*; Eggertsen, Jamie. "What You Need to Know: AI Disclosure Rules in Legal Filings." *Eve Legal*, March 19, 2025. https://www.eve.legal/blogs/what-you-need-to-know-ai-disclosure-rules-in-legal-filings (accessed March 15, 2026).
73. UK AI Safety Institute. (2024). "AI Safety Institute approach to evaluations." *Department for Science, Innovation and Technology*. https://www.gov.uk/government/publications/ai-safety-institute-approach-to-evaluations/ai-safety-institute-approach-to-evaluations (accessed March 15, 2026).

Chapter 7

1. COLLINS. "Spotify." We Are Collins. https://wearecollins.com/work/spotify/ (accessed March 16, 2026).
2. Breznican, Anthony. "Steven Spielberg talks about 'Jaws'—the greatest summer movie ever made." *Entertainment Weekly*, June 8, 2011. https://ew.com/article/2011/06/08/steven-spielberg-jaws-interview/ (accessed March 16, 2026).
3. "Canva, Our International Expansion: A Global Vision with a Local Mindset," December 10, 2024. https://www.canva.com/newsroom/news/international-expansion (accessed April 15, 2026).
4. "Canva Enterprise is now used by 95% of the Fortune 500." *Canva Newsroom*, 2025. https://www.canva.com/newsroom/news/one-year-canva-enterprise/ (accessed March 16, 2026).
5. "Introducing Canva Shield: Safe, fair, and secure AI." *Canva Newsroom*, November 5, 2025. https://www.canva.com/newsroom/news/safe-ai-canva-shield/ (accessed March 16, 2026).
6. "Introducing Canva's Creative Operating System." *Canva Newsroom*, October 29, 2025. https://www.canva.com/newsroom/news/creative-operating-system/ (accessed March 16, 2026).
7. Brynjolfsson, Erik; Daniel Rock; and Chad Syverson. "The Productivity J-Curve: How Intangibles Complement General Purpose Technologies," *American Economic Journal: Macroeconomics* 13, no. 1 (2021): 333–372. https://doi.org/10.1257/mac.20180386 (accessed June 17, 2026).

8. Bianconi, Eva; Allison Piovesan; Federica Facchin; Alina Beraudi, et al. "An Estimation of the Number of Cells in the Human Body," *Annals of Human Biology* 40, no. 6 (2013): 463–471. https://pubmed.ncbi.nlm.nih.gov/23829164/ (accessed March 16, 2026); "International Human Genome Sequencing Consortium Describes Finished Human Genome Sequence." National Human Genome Research Institute, October 20, 2004. https://www.genome.gov/12513430/2004-release-ihgsc-describes-finished-human-sequence (accessed March 16, 2026).
9. "International Human Genome Sequencing Consortium Describes Finished Human Genome Sequence." National Human Genome Research Institute.
10. Burley, Stephen K.; Helen M. Berman; Jose M. Duarte; Zukang Feng, et al. "Protein Data Bank: A Comprehensive Review of 3D Structure Holdings and Worldwide Utilization by Researchers, Educators, and Students," *Nucleic Acids Research* 50, no. D1 (2022): D437–D451. https://www.ncbi.nlm.nih.gov/pmc/articles/PMC9599165/ (accessed March 16, 2026).
11. Jumper, John; Richard Evans; Alexander Pritzel; Tim Green, et al. "Highly Accurate Protein Structure Prediction with AlphaFold," *Nature* 596 (2021): 583–589. https://www.nature.com/articles/s41586-021-03819-2 (accessed March 16, 2026).
12. "AlphaFold Protein Structure Database." DeepMind and EMBL-EBI. https://alphafold.ebi.ac.uk/ (accessed March 16, 2026).
13. Saplakoglu, Y. (2024). "How AI Revolutionized Protein Science, but Didn't End It." *Quanta Magazine*. https://www.quantamagazine.org/how-ai-revolutionized-protein-science-but-didnt-end-it-20240626/ (accessed March 16, 2026).
14. Tabula Sapiens Consortium. "The Tabula Sapiens: A Multiple-organ, Single-cell Transcriptomic Atlas of Humans," *Science* 376, no. 6594 (2022): eabl4896. https://pubmed.ncbi.nlm.nih.gov/35549404/ (accessed March 16, 2026).
15. Pearce, James D.; Sara E. Simmonds; Gita Mahmoudabadi; Lakshmi Krishnan, et al. "A Cross-Species Generative Cell Atlas Across 1.5 Billion Years of Evolution: The TranscriptFormer Single-cell Model," *bioRxiv* (2025). https://www.biorxiv.org/content/10.1101/2025.04.25.650731v1 (accessed March 16, 2026).

16. Bunne, Charlotte; Yusuf Roohani; Yanay Rosen; Emma Lundberg, et al. "How to Build the Virtual Cell with Artificial Intelligence: Priorities and Opportunities," *Cell* 187, no. 25 (2024): 7045–7063. https://www.cell.com/cell/fulltext/S0092-8674(24)01332-1 (accessed March 16, 2026).
17. Saplakoglu. "How AI Revolutionized Protein Science, but Didn't End It."
18. Horvitz, Eric. "When AI Meets Biology: Promise, Risk, and Responsibility," *Microsoft Research Blog*, October 6, 2025. https://www.microsoft.com/en-us/research/blog/when-ai-meets-biology-promise-risk-and-responsibility/ (accessed March 16, 2026).
19. "Framework for Nucleic Acid Synthesis Screening." Office of Science and Technology Policy, April 2024; revised September 2024. https://aspr.hhs.gov/S3/Documents/OSTP-Nucleic-Acid-Synthesis-Screening-Framework-Sep2024.pdf (accessed March 16, 2026).
20. "11 Fun Facts About Your Brain." Northwestern Medicine, https://www.nm.org/healthbeat/healthy-tips/11-fun-facts-about-your-brain (accessed March 16, 2026).
21. NVIDIA Corporation. NVIDIA H100 Tensor Core GPU Datasheet, 2023. https://www.nvidia.com/en-us/data-center/h100/ (accessed March 16, 2026).
22. Hamilton, Marc. "Chill Factor: NVIDIA Blackwell Platform Boosts Water Efficiency by Over 300x," *NVIDIA Blog*, April 22, 2025. https://blogs.nvidia.com/blog/blackwell-platform-water-efficiency-liquid-cooling-data-centers-ai-factories/ (accessed March 16, 2026).
23. "International Data: Electricity Consumption," U.S. Energy Information Administration (EIA). https://www.eia.gov/international/data/world/electricity/electricity-consumption (accessed March 16, 2026).
24. "Transmission Impact Assessment," *U.S. Department of Energy*, October 2024. https://www.energy.gov/sites/default/files/2024-10/DOE_OP_2024_Report-Transmission_Impact_Assessment.pdf (accessed March 16, 2026).
25. Maio, Pat. "After 15 Years Of Permitting, TransWest Wind Transmission Project Is Still 5 Years From Going Live," *Cowboy State Daily*, April 20, 2023. https://cowboystatedaily.com/2023/04/20/after-15-years-of-permitting-transwest-wind-transmission-project-is-still-5-years-from-going-live/ (accessed March 16, 2026).

26. "Why it's so hard to build new electrical transmission lines in the U.S.," *CNBC*, February 22, 2023. https://www.cnbc.com/2023/02/21/why-its-so-hard-to-build-new-electrical-transmission-lines-in-the-us.html (accessed March 16, 2026).
27. Cassauwers, Tom. "Machine Learning Can Boost the Value of Wind Energy." *DeepMind Blog*, February 18, 2019. https://deepmind.google/discover/blog/machine-learning-can-boost-the-value-of-wind-energy/ (accessed March 16, 2026).
28. Microsoft Azure Quantum. "How Microsoft's AI Screened 32 Million Candidates to Find a Better Battery." January 9, 2024. https://azure.microsoft.com/en-us/blog/quantum/2024/01/09/unlocking-a-new-era-for-scientific-discovery-with-ai/ (accessed March 16, 2026).
29. "Magnetic control of tokamak plasmas through deep reinforcement learning," *Nature*, February 16, 2022. https://www.nature.com/articles/s41586-021-04301-9 (accessed March 16, 2026).
30. EU Covenant of Mayors. "Stockholm, Sweden: Heat Recovery from Data Centres." *EU Covenant of Mayors*, October 10, 2023. https://eu-mayors.ec.europa.eu/en/news/stockholm-sweden-heat-recovery-data-centres (accessed March 16, 2026).
31. Smolaks, Max. "DCD>Energy Smart: Mayor of Stockholm Welcomes Data Center Projects." *Data Center Dynamics*, March 15, 2018. https://www.datacenterdynamics.com/en/news/dcdenergy-smart-mayor-of-stockholm-welcomes-data-center-projects/ (accessed March 16, 2026).
32. "Fortum and Microsoft announce world's largest collaboration to heat homes." *Fortum*, March 17, 2022. https://www.fortum.com/media/2022/03/fortum-and-microsoft-announce-worlds-largest-collaboration-heat-homes-services-and-businesses-sustainable-waste-heat-new-data-centre-region (accessed March 16, 2026).
33. "Exploring a space-based, scalable AI infrastructure system design," *Google Research Blog*, November 4, 2025. https://research.google/blog/exploring-a-space-based-scalable-ai-infrastructure-system-design/ (accessed March 16, 2026).
34. Feilden, Ezra; Adi Oltean; and Philip Johnston. "Why we should train AI in space," *Starcloud White Paper* (September 2024), https://starcloudinc.github.io/wp.pdf (accessed March 16, 2026).

35. Lee, Angie. "How Starcloud Is Bringing Data Centers to Outer Space," *NVIDIA Blog*, October 15, 2025. https://blogs.nvidia.com/blog/starcloud/ (accessed March 16, 2026).
36. Lee. "How Starcloud Is Bringing Data Centers to Outer Space."
37. "How Will Crusoe and Starcloud Build Data Centres in Space?" *Data Centre Magazine*, October 2025. https://datacentremagazine.com/news/how-will-crusoe-and-starcloud-build-data-centres-in-space (accessed March 16, 2026).
38. "China Is Building The First 'AI-Powered' Data Center In Space." *IFLScience*, October 7, 2025. https://www.iflscience.com/china-is-building-the-first-ai-powered-data-center-in-space-why-81060 (accessed March 16, 2026).
39. "Lonestar's Data Center is ready for the Moon." *PR Newswire*, February 10, 2025. https://www.prnewswire.com/news-releases/lonestars-data-center-is-ready-for-the-moon-intuitive-machines-and-lonestar-complete-final-installation-and-testing-of-lonestars-data-center-payload-302372021.html (accessed March 16, 2026).

Acknowledgments

This is a book about trust. It exists because of trust – the early, unwavering kind that many have given me through the years.

I started writing it while I was pregnant with my second child. I am finishing it with a nine-month-old on my hip and a two and a half-year-old who would rather I read *his* book. Between those two points: nearly a hundred interviews, thousands of pages of research, and the strange clarity of postpartum – when you sit still, stare at a small human, and finally confront the questions you have carried for years but never stopped long enough to answer. One would not leave me alone: What does trust mean when the technology outpaces the humans using it?

I could not have answered it without the people who made me ask it in the first place.

Stuart Russell's *Human Compatible* gave me the intellectual courage to question the trajectory of AI. Steve Quake showed me the staggering promise of what this technology becomes when wielded with rigor and conscience. Marc Benioff's vision and my years at Salesforce taught me that trust is a leadership decision. And my allies in digital trust – the builders, policymakers, researchers, and operators I have been in the trenches with over the past decade – you shaped how I think about what this technology can become and what it can cost. You planted seeds that became a book.

Lisa Shalett, Karl Ulrich, Matt McLarty, Catherine Connelly, Jennie Baird, Katherine von Jan, Leland Machmeyer, Mike Samson,

Debo Olaosebikan, Lee Redden, Phil Tomlinson, Angela Johnson, Ben Tseng, Woodson Martin, Ravi Viswanathan, Seth Matlins, Richard Socher – you are my brain trust. Some of you gave me the confidence to start. Some of you opened your networks so I could bias less, question more, and sink deeper. Some of you challenged assumptions, stress-tested arguments, and told me the truth when it would have been easier to nod. Some of you read drafts with ruthless honesty and unwavering conviction that this book mattered. Each of you did exactly what I needed, when I needed it. You made the ideas in this book braver and more honest than I could have made them alone. You made a solitary journey feel less lonely and worthwhile from the very start. This book exists because you believed in it before I did.

My fellow Young Global Leaders at the World Economic Forum – you brought global perspectives, cultural nuances, and what's actually happening on the ground across the world into this work. But you also showed up for the personal journey – the one where I was writing a book while raising two small children and wondering if I could do both. You were there intellectually and emotionally, and you reminded me why this work is urgent and why it cannot be done from any single vantage point.

To the entire team at Wiley – you trusted a first-time author with a big argument. In a book about trust, I do not take that lightly. I began this book asking how humans and machines can work together without losing what makes us human. That question shaped my writing process, too. AI assisted with research and editorial tasks; every word was reviewed, edited, and decided by me, peer reviewers, and my Wiley editors. Human agency isn't just this book's argument. It was its method.

Archie and Cali – you are too young to read this, but you are the reason it exists. Every chapter about agency, about what we preserve and what we surrender, was written with a simple hope: that you will grow up thinking for yourselves, questioning what sounds authoritative, and never outsourcing your judgment just because a machine makes it easier. That is what this book is for. That is what your mother is for.

Dan – you held everything together so I could fall apart into these pages. You challenged every lazy argument and forgave every late night. You are the unwavering constant in a life I have spent in motion.

Mom and Dad – I have traveled far from you for 20 years. Across oceans, continents, and life choices you rarely agreed with. You have never fully understood why I couldn't just stay closer to home. But you have always, eventually, trusted me. That may be the most extraordinary act of trust in this entire book.

After all, trust is the leap of faith that becomes accomplishment, innovation, and impact. It is the most beautiful thing one can experience, and the luckiest thing one can receive. Every name on this page is someone who gave me theirs. This book is my case for why we must protect it.

About the Author

Tiffany Xingyu Wang is a global leader in digital trust and AI safety who has spent the past decade building systems that protect over a billion users online. Named a World Economic Forum Young Global Leader and Forbes' Most Entrepreneurial CMO, she serves as a non-resident Senior Fellow at the Atlantic Council and a member of the World Economic Forum's Frontier AI Systems and Capabilities group.

Index

Note: Page numbers in *italics* refer to figures.